The Permanent Campaign

Steve Jones
General Editor

Vol. 81

The Digital Formations series is part of the Peter Lang Media and Communication list.
Every volume is peer reviewed and meets
the highest quality standards for content and production.

PETER LANG
New York • Washington, D.C./Baltimore • Bern
Frankfurt • Berlin • Brussels • Vienna • Oxford

Greg Elmer
Ganaele Langlois
Fenwick McKelvey

The Permanent Campaign

New Media, New Politics

PETER LANG
New York • Washington, D.C./Baltimore • Bern
Frankfurt • Berlin • Brussels • Vienna • Oxford

Library of Congress Cataloging-in-Publication Data

Elmer, Greg.
The permanent campaign: new media, new politics /
Greg Elmer, Ganaele Langlois, Fenwick McKelvey.
p. cm. — (Digital formations; v. 81)
Includes bibliographical references and index.
1. Communication in politics. 2. Social media—Political aspects.
I. Langlois, Ganaele. II. McKelvey, Fenwick. III. Title.
JA85.E46 324.70973—dc23 2012034607
ISBN 978-1-4331-1606-3 (hardcover)
ISBN 978-1-4331-1593-6 (paperback)
ISBN 978-1-4539-0920-1 (e-book)
ISSN 1526-3169

Bibliographic information published by **Die Deutsche Nationalbibliothek**.
Die Deutsche Nationalbibliothek lists this publication in the "Deutsche Nationalbibliografie"; detailed bibliographic data is available on the Internet at http://dnb.d-nb.de/.

The paper in this book meets the guidelines for permanence and durability of the Committee on Production Guidelines for Book Longevity of the Council of Library Resources.

29 Broadway, 18th floor, New York, NY 10006
www.peterlang.com

Printed in the United States of America

Contents

Acknowledgments

The authors would like to thank our previous collaborators at the Infoscape Research Centre at Ryerson University, without whom this project would never have seen the light of day: Zachary Devereaux, Elley Prior, Peter Ryan, A. Brady Curlew, Joanna Redden, Mark Brosens, Isabel Pedersen, and Paul Goodrick. Our research centre's programmers and designers have also played a key role in our capacity to conduct research. We are particularly grateful to Many Ayromlou, Paul Vet, Rob King, Greg J. Smith, and Patricio Davila.

Portions of this book were also made possible through a number of collaborative research projects, invited lectures, and fellowships. We would like to thank Rob Proctor, Yu Wei Lin, and in particular Rachel Gibson for the opportunity to share—and receive tremendous feedback—for our work at the University of Manchester. Similarly, Nick Jankowski, Steven Sams, and in particular Maurice Vergeer provided very helpful feedback for early versions of Chapter Five. Kim Sawchuk, Theo Rohle, and Oliver Leistert provided crucial encouragement and feedback on Chapter Three. Mike Thelwall offered a wonderful opportunity to seek feedback from researchers at his centre at the University of Wolverhampton. Richard Rogers and students at the Digital Methods Initiative (DMI) at the University of Amsterdam invited us to present and further develop an early version of Chapter Six. Susan Ormiston, Paul Hambleton, and Allison Brachman were instrumental in facilitating and initially broadcasting our research from Chapter Five through various media platforms at the Canadian Broadcasting Corporation (CBC News Division).

A fellowship at the Cultures of the Digital Economy (CoDE) Research Institute at Anglia Ruskin University was instrumental in providing time to write the book's first chapters. The authors would like to thank Jussi Parikka, Joss Hands, and David Skinner (Anglia Ruskin University) for their feedback on previous versions of many of our chapters. Similarly, the work of Andrew Chadwick, Kirsten Foot, Steven Schneider, and Philip Howard provided important conceptual insights during the writing of this book. Charo Lacalle at the Autonomous University of Barcelona, Tristan Thielmann at the University of Siegen, and Helmut Winkler at the University of Paderborn also graciously invited us to present portions of our book.

The authors would like to thank Erika Biddle (York University) for her valiant efforts to polish up our prose and grammar, and for laying out this book. Similarly, we wish to thank Steve Jones and Mary Savigar for their patience and feedback on our manuscript.

This project was funded by a number of sources, mainly the Social Science and Humanities Research Council (Canada) and the Bell Globemedia Research Chair, Ryerson University. Funds were made available through a contractual partnership with the Canadian Broadcasting Corporation (CBC News) and the Korean government's World Class University (WCU) project.

Portions of Chapter Four were published in the *Canadian Journal of Communication* 34 (3). A version of Chapter Five will be published in an upcoming issue of *New Media & Society*. And, an earlier version of Chapter Six will appear in an upcoming issue of *Information Polity*.

Greg Elmer, Ganaele Langlois, and Fenwick McKelvey
Toronto, May 2012

CHAPTER 1

The Permanent Campaign: Platforms, Actors, and Objects

After leaving his position as press secretary for George W. Bush's White House, Scott McClellan spent some months in the political wilderness before reappearing on the national stage with a tell-all book and a speaking tour. A solid Bush loyalist for close to three years, McClellan re-entered political life to denounce not only his old political boss but also the permanent campaign in Washington: the hyper-partisan, insider-driven political game. A similar story played out north of the border in Canada, where Tom Flanagan, former advisor to Prime Minister Stephen Harper, distanced himself from his former boss' permanent-campaigning mode, one that he argued had turned the Conservatives into a "garrison party":

> Just as chronic warfare produces a garrison state, permanent campaigning has caused the Conservative Party to merge with the campaign team, producing a garrison party. The party is today, for all intents and purposes, a campaign organization focused on being ready for and winning the next election, whenever it may come (MacCharles, 2010).

While such high-profile political operatives have invoked the "permanent campaign" in an effort to raise questions about recent partisan wranglings, the concept was first introduced by Patrick H. Caddell, an advisor to then-President-elect Jimmy Carter. According to *Time* magazine, Caddell told Carter in 1976: "it is my thesis governing with public approval requires a continuing political campaign" (Klein, 2005). Caddell's remark came at a particularly contested moment in American political history, one that witnessed a renewed sense of patriotism and partisanship at the end of a decade defined by a loss of public confidence in political leadership (i.e., the post-Vietnam and Watergate years). In such a politically volatile climate, where politicians came under close scrutiny from not only their political opponents but also members of the media and voters, Caddell's report was among the very first to recommend bridging the tactics, time, and technologies of governance and electoral campaigning. It has been argued, however, that permanent campaigning has a much longer history. Some of the earliest efforts to stage-manage mass media events occurred at "whistle stops" on presidential campaign trains in the United States (Ornstein & Mann, 2000). Ornstein and Mann thus characterize permanent campaigning as a form of "technical management" of the political sphere; or, as an effort on the part of the political class (consisting of elected representatives and their assorted operatives) to strictly control the terrain and terms of political life (p. 38).

As an object of study, the permanent campaign has also effectively brought together a group of American political scientists in search of empirical methods and approaches to the study of partisan governance. Murray and Howard (2002), for example, argue that a politician's travel and public appearances, particularly at public events, serve as an indicator of permanent campaigning. Likewise, Doherty's (2007) study of non-election-year presidential travel indicates a pattern of visiting "large competitive states." The widespread growth of political-opinion polling by governments worldwide also points to an ongoing concern with electability, even after a successful electoral campaign (Murray & Howard, 2002).

While we should be careful to note distinguishing factors across national case studies, permanent campaigning as emergent 24/7 cycles of partisan forms of publicity typically share common modernist traits. In the political sphere, these result from: (1) the bureaucratization of political parties; (2) the emergence of formalized/professionalized election apparatuses (or in the vernacular, party "machines"); and (3) executive-dominated political cultures where "backbencher" legislators are rewarded for their loyalty and harshly punished for dissent from the party line and party votes. Election cycles also play a significant role in mediating the ebbs and flows of partisan and political life across jurisdictions (Alesina & Roubini, 1992). The splintering of electorates—through the growth of regional, linguistic, anti-immigrant, and protest parties—also has cultivated permanent campaigning, as unstable minority coalition governments and "hung" parliaments increase the frequency or at least the possibility of shorter election cycles.

Lastly, the emergence of permanent campaigning can also be tied to the exponential rise in political advertising and fund-raising—much of which is accomplished by appealing to the more ardent and ideologically committed members of particular political parties. Similarly, governments, opposition parties and other minority parties now routinely engage in legislative setups, gotcha-politics, and scheduled votes on wedge issues that attempt to draw sharp policy distinctions between parties. Such distinctions typically touch upon particularly contentious or "red meat" social issues such as abortion, gun control, and immigration rights (Hillygus & Shields, 2008).

Whatever the emphasis, each of these scholars explicitly or implicitly raises the question of partisanship and permanent campaigning out of a concern for the conduct of political life; in short, for ethical reasons. Permanent campaigning suggests a waning of the central tenets of political representation, whereupon elected representatives, particularly members of the ruling party, displace their responsibility for the common good—for all citizens—in favor of the technical management of continued electoral success. The perils of permanent campaigning lie in the privileging of partisanship over governance. Ironically, this perilous state of affairs is both constructed and destabilized by the very same set of factors; namely, by the rapid introduction of information and communication technologies into political life. Ornstein and Mann (2002) likewise make this important point in taking us back to the first railway-enabled political campaigns. They discuss permanent campaigning primarily as a mediated phenomenon—as a process of gaining greater control over political messages, particularly mass-mediated or reported events, "staged" or otherwise (see also Marvin, 1988).

For decades, network news has served as a key site of political debate and political campaigning, be it through its coverage of events, talk shows, or televised candidate debates. Despite the predictability of such coverage, in terms of its discrete broadcast timing and scheduling, television news was fundamentally altered in 1980 with the introduction of the Cable News Network (CNN). Similar networks around the globe soon joined CNN in introducing the 24-hour news cycle to political life. The subsequent need for more hours of news coverage greatly affected political programming on television, leading to the development of an expanded class of political pundits—typically, former members of the political or journalistic class—whose newfound freedom from professional conventions led to the airing of unabashed partisan sloganeering and venomous debate. The multiplication of spaces for political debate created the need for more political spokespeople, thereby valorizing communicators over legislators, orators, or parliamentarians (Cushion & Lewis, 2010; Dagnes, 2010).

Building upon these changes in news cycles and political communications, this book investigates how the introduction of the Internet and the World Wide Web has redefined and otherwise contributed to a new paradigm of permanent campaigning. Moving beyond discussions of expanded media time, reporting and commentary, or the timing and coverage of public affairs, or even the political tone of news coverage, this book develops an understanding of, and new perspectives on, methods and theories of networked political communication. These perspectives are based upon what we see as a socially mediated and distributed battle over political opportunities presented by the Web—one that has greatly complicated the relatively fixed roles of journalist, citizen, and politician discussed by Ornstein and Mann (2002). The political establishment, and to a lesser extent the mainstream press, are now faced with a destabilized set of relationships produced by distributed and interactive forms of communication and networking; in other words, by techniques and technologies that afford two-way communication in real time. Contemporary forms of permanent campaigning thus continue to invoke partisan tactics and talking points (centralized message control) in a political landscape that deploys campaign-style tactics in the routine running of government. However, so-called "participatory media" (see Jenkins, 2006)—namely, Web 2.0's networked platforms (e.g., blogs, microblogs, online videos, and social networking)—have also challenged centralized and hierarchical forms of political governance and campaign management. Permanent campaigning today is largely defined by the ongoing attempts by various political actors to harness and otherwise manage the opportunities wrought by new information and communication technologies, including media and voter-management technologies (Howard, 2006).

Without the rigid definition afforded by traditional political roles and institutional hierarchies, contemporary permanent campaigning clearly contributes to a murkier political sphere—a blurring of temporalities, rules, conventions, and professional roles that have produced new hybrid actors such as "citizen-journalists" and networked spaces for official government policy announcements (e.g., Twitter). Even electoral results and ensuing claims of political "mandates" have been remediated through this cloudy political landscape. On this count, the 2000 U.S. presidential election marked a pivotal moment in contemporary Western political history. Online activists, fundraisers, and pundits engaged in a protracted campaign to settle the terms of the incomplete Electoral College results. While the uncertainty of who won the 2000 U.S. presidential election underscored a crisis in American political opinion, it also served to highlight the role of technology in elections—the abuse of and mistrust in voting machines, and of course, the use of Internet sites to debate and spread questionable "facts" during the ensuing constitutional crisis. More recently, similar post-electoral political dynamics have occurred in Canada and the United Kingdom, when established political parties failed to secure a majority of seats in their respective Houses of Commons.

Platforms, Subjects, and Objects

Theoretical, conceptual, and empirical studies of the socially mediated aspects of permanent campaigning must recognize not only the speed, hyper-immediacy, and always-on ("24/7") dimensions and potential of modern political communications, but also the speed at which Web platforms, political roles, and forms of communication themselves constantly change. In other words, networked permanent campaigning is as much about flexibility, adaptability, and adoption as it is about partisan politics. In this book we argue that networked permanent campaigning intersects with—that is, seeks to manage and is itself managed by—three constantly shifting phenomena: (1) the spaces of communication and campaigning itself (particularly social media platforms); (2) partisan participation, action, and subjectivity (roles of political actors); and lastly, (3) the digital encoding and circulation of political communications (or "issue-objects").

This introductory chapter first questions how partisan life is enacted upon and across Web-based platforms. Sites such as YouTube, Facebook, and Twitter contain their own unique set of practices, rules, and networking opportunities. By understanding how such platforms interact and are themselves networked—in terms of how one can both distribute content across them, and also view and interact with comments from a series of such platforms—we can begin to understand this new terrain of networked politics.

Second, political campaigning during "off-peak" hours or political crises calls into question the very nature of what it means to be a political partisan in the networked age. This focus on partisan subjectivity, however, does not simply seek to understand the blurring of political roles among those interested in and active within political circles (officially or otherwise). Rather, it focuses on the role that such new actors play in heightening and privileging continuous partisan and political activity as a central characteristic of contemporary permanent campaigning. Indeed, it logically flows from our first point, above—as was the case in the development of the 24-hour news cycle, more opportunities, spaces, and sites for political communication create the need for new political "staff": individuals who can be called upon to contribute to debates, dialogues, and other forms of political communication that support political parties, including their elected leaders and public policies.

Lastly, in this chapter we analyze the tactical and strategic deployment of [illegible]bjects for political goals as the primary mode of political campaigning in [illegible]rked landscape. In this concluding section, we investigate how discrete [illegible] (videos, blog posts, digital images, graphics, hyperlinks, etc.) are used to unite political campaigns across platforms (see Foot & Schneider, 2006, pp. xxii, 263), while also serving as "stand-alone" object-signifiers—artifacts that connote specific political meaning (i.e., political issues, positions, ideologies, etc.) in and of themselves. That is how they invoke and seek to influence and frame specific political issues and debates.

1. The Platform of Permanent Campaigning

The concept of the permanent campaign is inseparable from the development of media systems. The transition from mass media to networked media, however, has radically changed the communicative terrain upon which forms of political campaigns are shaped. Networked media fundamentally challenges long-held assumptions about media reception, participation, and political roles. The Web's distributed networked environment therefore forces us to reconceptualize and redefine the very term "political communication" as one that must now account for the ever-expanding capacity for information storage and retrieval, multiple entry points of communication, and expanded sites and modes of self-expression. Current buzzwords used to describe online communication—for example, "Web 2.0," "social media," and "participatory culture"—all try to express shifting possibilities for political communication on distributed networks. In this new context, the boundaries between producers and receivers are constantly blurred, with seemingly no limits to the expansion and circulation of information and content. The traditional gatekeeping role of the mainstream media is under threat from participatory communication platforms that host grassroots and independently produced content. Whereas previous forms of permanent campaigning focused on enclosing the flows of communication, we see the main challenge for political communication today as one of managing seemingly open-ended flows of communication. Contemporary networked media is an information geography that affords a multiplicity of sites, spaces, and routes of political communication.

While open-ended communication is a
contemporary networked media, this does not i
limits. The hyperlink—the original structuring p
Web—formerly served as both a protocol of com
for organizing information across Web docum
interest. As such, methodological approaches
and subsequent flows of information online

hyperlinked hubs and patterns across Web pages and sites (Park & Thelwall, 2003). The current structure of the Web is also marked, however, by the rise of the operating software platform model (Gillespie, 2010; Langlois, Elmer, McKelvey & Devereaux, 2009). Web 2.0 platforms such as YouTube and Facebook have emerged as largely enclosed and self-regulated operating systems that operate "on top of" existing Web protocols. As opposed to a hyperlink-defined Web 1.0 framework, Web 2.0 platforms manage content and users through a combination of complex algorithms and protocols. The common goal of most if not all participatory platforms is to accommodate (in terms of easy-to-use publishing tools and expansive databases) large amounts of user-generated content. In the context of networked politics, content may be created by a diverse range of users, from official party staff to interest groups and citizens. What distinguishes one platform from another is the form taken by this content: for example, videos on YouTube, digital pictures on Flickr, short posts on Twitter, and more multimodal approaches on social networking sites such as Facebook.

Each Web 2.0 platform also differs from others in terms of the ways its users can relate to content—the kinds of comments they can make, the possibility of exporting content from one platform to another, etc. While such interactive functions are often clearly highlighted on platform interfaces for users, the main distinguishing feature among participatory platforms concerns the ranking of information. It is not enough for participatory platforms to offer databases or repositories for various forms of information and communication; the participatory platform must also offer specific logics by which information can be meaningfully retrieved, ranked, and circulated—in short, made visible. Such logics differ from one platform to the next, and require different arrangements and practices between software and users, thus posing significant problems for those seeking to orchestrate coherent and cohesive political communications and campaigns across platforms. For instance, some platforms might push software-produced recommendations onto users. In this case, the software is in charge of extracting the meaningful information; that is, the kind of information that might be relevant to all users, or specific categories of users, or individual users. For example, YouTube currently offers two ways of accessing videos: a general search box that ranks results in terms of popularity (most-viewed videos), and personalized recommendations based on an individual user's viewing history. The search method is based on the logic that what is the most popular is also the most universally relevant, while the personalized recommendations are based on a user's specific patterns of viewership. Other platforms are more focused on enabling users to establish threads of relevant content through user-generated tags and keywords, or on allowing users to create their own feeds for following other user-generated content streams, as is the case with Twitter. Often, participatory platforms will multiply the different logics of retrieving and ranking content according to multiple definitions of what constitutes meaningful content.

Given the varying approaches to ranking and otherwise making content visible on Web 2.0 platforms, permanent campaigning on the Net can be understood as an attempt to manage the open-endedness of communication by maximizing the potential interface time, visibility, and ranking of specific digital objects—be they text, video, images, or audio clips (see Chapter Five). Strategic interventions in the flow of open-ended content subsequently require enlisting online partisans to participate in political debate and organizing. Barack Obama's 2008 presidential campaign, for instance, was considered an Internet success in large part because it mobilized users to circulate messages about the campaign on multiple platforms—mostly via social networks, but also video sites and blogs. However, permanent campaigning is not characterized solely by an expanded sphere of political actors, as the Obama campaign reminds us; it also sees a concerted effort to sustain specific discourses, sound bites, and images across a number of media sites and over as long a period of time as possible. To achieve such effects requires an understanding and strategic use of the communicational logics within and across platforms that organize information according to a specific definition of meaningfulness.

It is often difficult to determine exactly how specific Web 2.0 platform logics of search, retrieval, and ranking work, because such processes are generally hidden from users, programmers, and others for proprietary reasons. Moreover, given the overall unique proprietary software code that governs respective Web 2.0 platforms, political networking and communicating on the Web involves constantly shifting sets of rules among 2.0 platforms that govern what can circulate across such sites. The circulation of information across platforms is increasingly enacted by individual partisans and online party supporters who "share" or recommend objects with other users, social networks, interfaces, and social media platforms, for example via Facebook "like" or Twitter RT or favorite buttons. While these modes of sharing simplify the circulation of content from the user's perspective, they also make it much more difficult to track patterns of circulation across platforms. Thus participatory platforms not only structure information, they also manage the circulation of information within and across platforms. This takes the form of controlling the conditions under which third parties—for example, political parties or the mainstream media—can access, retrieve, and customize the information made available on participatory platforms.

Another important distinguishing characteristic of participatory platforms, then, as opposed to the hyperlink-web 1.0 framework, is the means by which third parties access or mine large data sets on social media platforms—be it user-generated content or personal information about users (demographics, surfing behaviors, content preferences, etc.). Access to such information is typically enabled by so-called application programming interfaces (APIs), data intensive portals on social media or Web 2.0 platforms that allow third-party customized software programs to link to and obtain information from the software in charge of managing information on the participatory platform side. APIs can range from database search/retrieval software modules to more complex modules designed to provide users with value-added—typically proprietorial—content (user demographics and behavior). The YouTube API, for example, makes it possible to obtain information about most-viewed videos (e.g., number of views, date posted, and tags). The Facebook API allows for more complex applications, such as campaign applications that can require users to allow the application's owner to access their personal data and network of friends. Obama's 2008 presidential campaign, for example, launched its "Yes We Can" application using Facebook's API. The application made it possible for users to embed the "Yes We Can" music video onto their Facebook page and also helped gather data about the geographical location and other networked demographics of potential supporters and financial contributors.

To summarize then, the open-ended field of communication—with far fewer central nodes and identifiable gatekeepers—that characterizes the participatory model offers a new terrain for permanent campaigning; one that is organized around platforms, each developing their own logic by which some content is rendered more visible or socially networked. The strategies of permanent campaigning are now focused on making use of these logics of visibility, with regards to both the mobilization and deployment of new actors and the development of specific back-end software processes that intervene in and obtain information about users via user-created content and user profiles.

2. New Political Actors: The Activation of Mobile Partisans

The history of political campaigning is one of expanding opportunities to engage in political communication through new sites, spaces, and times. While opportunities were once clearly defined in terms of formats, times, and roles, today we see a blurring of such distinctions with accompanying claims of a "participatory" culture on new media platforms and technologies that pave the way for political engagement (Jenkins, 2006). However, such participation is an intensely contested zone (see Scholz, 2006) which some have dubbed "politics 2.0" (Chadwick & Howard, 2008). This zone is defined by its relationships to existing forms of political power, even as it opens the field to a new plane of political life.

Politics 2.0 presents opportunities for nontraditional political actors to engage established actors and institutions, while at the same time enabling political parties to reach out and interact with these new political actors to staff the permanent campaign. Unlike in the past, these partisan individuals are not contracted or employed as staff of political parties, nor are they simply volunteers or grassroots members of political parties. With the introduction of 2.0 platforms—particularly those that host, serve, and otherwise facilitate the use of "blogs"—individuals interested in engaging with the political process can in near real-time publish their partisan opinions.

Before we turn to an analysis of the digital objects and discourses circulating across politics 2.0, we should introduce the core of any permanent campaign's mode of governance: the intensified and expanded field of the contemporary partisan. In examining partisanship, which has been the subject of intense criticism from radical democratic theorists such as Esposito (2005), the late conservative political theorist Carl Schmitt presents a particularly helpful point of departure. In *The Theory of the Partisan* (2007[1963]), Schmitt offers a rather institutionally minded definition of partisanship, one that invokes military metaphors into the civil sphere: "The partisan fights at a political front, and precisely the political character of his acts restores the original meaning of the word partisan. The word derives from party, and refers to the tie to a fighting, belligerent or politically active party or group" (p. 15). Yet, it's the more mechanically or technologically enabled form of partisanship that Schmitt offers to our 2.0 perspective, one that helps to understand the elastic relationship between partisan bloggers, for example, and political parties. Of the partisan, Schmitt writes:

> His mobility is increased by his motorization to such an extent that he is in danger of becoming completely disoriented. In the situations of the Cold War, he becomes a technician of the invisible struggle, a saboteur, and a spy. Already during World War II, there were sabotage troops with partisan training. Such a motorized partisan loses his telluric character and becomes only the transportable and exchangeable tool of a powerful central agency of world politics, which deploys him in overt and covert war, and deactivates him as the situation demands (p. 22).

What Schmitt's "mobile or motorized partisan" offers, then, is the figure of a political actor motivated by partisan political goals, though not necessarily tied (as a matter of routine or convention) to the everyday machinations of political parties. While such actors exhibit nomadic traits—taking from Schmitt's definition of partisanship—they also maintain a semblance of party loyalty, which he defines as one initiated or activated by political parties. This is the calling to arms of the mobile partisans for the good of the party. Bloggers, Twittering politicos, YouTubers, or vloggers and the like, are both proactively and subtly courted by individuals, factions, or campaigns from political parties. The permanent campaign requires reliable staffing. Thus, online political activists are consistently monitored by political parties, through their social media monitoring services, to determine who among the partisan bloggers can be trusted when the political stakes are raised. Of course, the notion of the nomadic, untethered 2.0 actor also highlights—and some might argue exacerbates—pre-existing and just-below-the-surface factions within political parties themselves. The ability of bloggers, for example, to anonymously disseminate and otherwise promote insider political information is consistently used by party members, leaders, and factions to disrupt internal agendas and embarrass political personalities (Flanagan, 2009, pp. 239–240). Indeed, leadership crises and contests within political parties often exhibit the most intense and mortal coupling of online partisans and established political masters.

Through such party-based connections, partisan bloggers and other online activists establish ongoing relationships with staff and representatives of the political parties and their various proxies. The importance of reputation, and the ability to consistently deliver insider information from political masters and to anonymously leak political information out of party offices, continues to cultivate a symbiotic relationship between individual bloggers and other online partisans and official staff from within established political parties. Such relationships offer a myriad of ex-distanced or otherwise virtual forms of political maneuvering that were previously restricted to selective and sometimes surreptitiously under-the-door leaks to members of the mainstream media.

The rise of such amateur partisan actors has of course provided the media with compelling figures to serve as Internet-based, user-generated savvy members of the technorati. The attempt to provide readers and viewers with commentary from technologically savvy and politically verbose newcomers is only one component in this newly reconstituted sphere of political punditry and posturing. The media—as well as political parties and institutions—call upon bloggers to enter the mainstream political discourse for strategic reasons. For the mainstream media, the growth of political consultancy, or professional talking-head punditry, has become stale—itself undermined by a near-monopolistic sphere of mediated political discourse. The tethering of such amateur political actors then mixes the *vox populi*, or "streeters" as the media refer to them (interviews with citizens typically walking to or from work or school), with more politically informed punditry. Since these bloggers do not officially speak for political parties, the media can routinely turn to them in the aggregate—e.g., the conservative or "Republican" blogosphere—any time of the day or night to obtain voices of the partisan political mood.

3. Issue-Objects: Discrete and Networked Tactics in the Permanent Campaign

The motivations and agendas of political partisans have always played a key role in defining the terrain of political conflict. In the networked age we see such partisans as intensifying the ongoing shift from traditional "iron triangle" arrangements (connections between elected officials, interest groups, and the bureaucracy) to what Hugh Heclo describes as *issue networks*. In his analysis of President Carter's administration, Heclo remarked that these "stable sets of participants coalesced to control fairly narrow public programs" had given way to issue networks that "comprise a large number of participants with quite variable degrees of mutual commitment or of dependence on others in their environment" (Heclo, 1978, p. 102). The content of politics is as broad as the term *issue*—a sort of catch-all category to define all the scandals, crises, perspectives, and concerns that spark entry into the political fray. Once they've established residence in the fray—through party manipulation, personal interest, or both—the partisan fights to have their message heard amidst all the political noise.

Expression in the permanent campaign entails a calculated maneuver to construct issues in order to gain scarce attention. Attention refers to both the prominence of issues in current political agendas and in the memory of the event itself (Lazzarato, 2006). Actors seek not only to bring their concerns to the forefront, but also to ensure their version of an event remains the preferred interpretation. In doing so, partisans attempt to construct issues in the most favorable light possible for their agenda. Capturing attention aligns the expression of politics with the calculated campaign style of talking points, stump speeches, and populist rhetoric. In each case, issues are carefully constructed in ways expedient to partisan and party goals and strategies. Partisans not only generate messages but also play a pivotal role in echoing and amplifying messages.

The permanent campaign is an "attention economy," a competition over the scarce resource of political attention (Lanham, 2007). Too often, political issues and concerns are ignored or fall on deaf ears. Bloggers, as Geert Lovink (2008, pp. xxviii, 312) points out, fear nothing more than zero comments—the sign that perhaps no one has read their post. Successful actors in the permanent campaign avoid the dreaded zero comments or, alternatively, cause an opponent's message to fall on deaf ears. As Elmer E. Schattschneider writes, "Some issues are organized into politics while others are organized out" (1960, p. 71). Political debate, in other words, hinges on capturing attention in the campaign—to win recognition or, at least, cause an opponent to be discredited.

Making an impact in the permanent campaign by capturing and sustaining attention is centered on discrete digital objects. Permanent campaigning, in other words, draws upon an object-oriented form of politics (Callon, Lascoumes & Barthe, 2009; Latour, 2005; Marres, 2005). Amid the clutter and clatter of political discourse, discrete digital objects (e.g., links, blog posts, YouTube videos, and Flickr photos) not only capture and embody contemporary politics, they *orient* the field of political communication and the means of publicity toward their sustained (meaning visible) circulation. Political missteps, consequently, are increasingly defined as political objects. Consider the racial slur uttered by Republican senator George Allen at a campaign rally in Virginia during the 2006 U.S. Senate election. Although the actual slur was made before a stage full of supporters, a Democratic campaign staffer recorded the incumbent senator's words on video. The Democratic campaign recognized the potential of the video and worked to circulate a media object that could discredit their opponent. They uploaded the video to YouTube, and then circulated the object to the mainstream media and partisan operatives. Bloggers and news sites re-circulated the object by embedding it in their Web pages.

An object-oriented politics has increasingly intensified during the past few tumultuous years in Canadian politics. The 2008 coalition crisis—which saw Canada's governor general settle a potential constitutional crisis—stands out as a major example of how political parties and their supporters leverage digital media objects. Just a few months after a federal election, Canada's Conservative government introduced what many anticipated to be a routine economic update. The announcement, however, contained what almost all political commentators viewed as a poison pill: legislation to remove public funding of federal political parties. The move directly threatened to cripple the indebted opposition parties, who emerged within days to announce their intention to join forces to defeat the government and replace them in the House of Commons with a rare coalition government. To preempt the defeat of the governing Conservatives, the Canadian prime minister abruptly ended the parliamentary session in a controversial parliamentary maneuver, in effect staving off defeat at the hands of the coalition-government-in-waiting.

With the closure of Parliament, the Internet lit up with campaigns for and against the proposed coalition government. Facebook groups—framed as either "I'm part of the 62 percent majority" (invoking the percentage of votes received by all opposition parties) or an "undemocratic coalition"—emerged for each side. While such campaigns were hosted on the popular social networking platform, it was typically specific blog posts—or objects—that served to highlight the content and further activities for both sides on the issue, including protests in city streets. The "coalition crisis," as it came to be known, thus demonstrated the multidimensional, multimediated life of a political object. Objects exist in a variety of media—such as the echo of a slogan in a partisan blog post or the talking points used on a debate show. These expressions calcify to the object, fixing characteristics that define an object's trajectory and significance. These expressions include not only content, but also modes of action, such as signing a petition or circulating a video. Studying the object involves digging into its strata to understand the object's formation and trajectory.

This object-oriented approach (see also Chapter Six) is also useful for understanding the dynamics of Barack Obama's "Yes We Can" campaign. The campaign deployed a range of multidimensional Web objects (e.g., official texts, videos and pictures, citizen-generated video responses, critiques and parodies), and importantly, links that function as deictics or pointers (Elmer, 2006) to different platforms (e.g., the official campaign website, the Facebook page, YouTube videos, fund-raising pages, organizing and get-out-the-vote sites, etc.). Various "Yes We Can" Web buttons (embedded in individual Facebook pages, blogs, and websites) facilitated political action and organizing, thereby transcending their use as mere declarations of allegiance and voting intentions. As an application—especially as a Facebook application developed by Obama campaigners—the online campaign also served as a covert polling measure to acquire information on supporters and would-be voters. As such, "Yes We Can" was a multilevel "traffic tag" (see again Chapter Six)—a digital title or tag that enables the wide circulation of Web objects while also creating a feedback loop from likely supporters and online organizers. From the user's point of view, "Yes We Can" served as both political content and a deictic pointer to a broader community of like-minded individuals.

The case of Obama's "Yes We Can" campaign illustrates a second important characteristic of an object: its circulation and ability to create networked ideas and technological and communicative affordances. Objects remain in motion, shifting, entangling, stabilizing, and evaporating. Some objects successfully circulate across the Web, such as a viral video, while others simply never make an impact, such as a blog post with zero comments. Circulation is a particularly significant process in a networked campaign. Circulation has intensified, in part due to the need to repeat a message to capture attention with digital systems that encourage the sharing of information.

Political campaigns depend on managing the circulation of content—not only attempting to promote objects (with the hopes of them going viral), but also remixing (re-editing) an object to change its original intended purpose. Similarly, leaking a potentially damaging story to a "friendly" blog helps shape it as an object before potentially negative articulations emerge. Circulation also includes hijacking an object. The candidate's name, for example, has become a key search term and point of entry into politics. Since interested voters conduct online searches for parties, platforms, and specific issues, campaigns routinely consider how best to position their candidate on major information aggregators such as Google. Online partisans and party operatives routinely seek to manipulate the search engine rankings so that negative critiques of their opponents appear at the top of search engine results. Additionally, politicians of all stripes buy Google ads to help direct to their campaign websites voters who search specific terms—or conversely, in order to direct voters to unflattering information about their political opponents. In France, for example, the political party Union pour un Mouvement Populaire (UMP) was under fire over allegations that their candidate Éric Woerth had received cash contributions from Liliane Bettencourt, heir to the L'Oréal cosmetics and beauty empire, that far exceeded the legal limit of €7,500 per year. The party bought ads for searches of "Bettencourt" to ensure their positive message entangled with returns around the controversy (Cario, 2010).

While the circulation of objects occurs in the context of strategic campaign communications, at times it evolves and devolves into an uncontrolled viral meme, an unintended consequence or accident. The management of political campaigning and its objects is a precarious one with constantly shifting algorithmic rules on social media platforms and information aggregators. An inability to master the political rules of the game, and the constant rewriting of the operating code used to enable social networking on the Internet, thus necessitates an ongoing rethinking of strategic political communication. This means that constant vigilance, not only to communication strategies but also to the very conditions of publicity, is required: in short, a permanent campaign.

Conclusions

This introductory chapter has highlighted three interconnected components of contemporary technologically mediated and networked political life: actors, platforms, and objects, the sum of which has cultivated an expanded political clock, an intensified form of partisanship that we refer to as the permanent campaign. New political actors (such as bloggers) have emerged as adjuncts to established institutional actors (such as political reporters, party and legislative staff, elected representatives, etc.). These mobile partisans, however, are unlike party workers, or even campaign volunteers who work primarily within party structures and conventions. As self-promotional subjects enacting an object-oriented form of politics, these online actors are proven network communicators. They know how to sustain interest for their objects (e.g., blog posts) through the use of network conventions such as hashtags, keywords, and strategies of embedding objects, and by manipulating information aggregators.

Knowledge of social media conventions provides for the necessary skills in a permanent campaign. The ability to easily publish, re-edit, comment, and circulate networked political content 24/7 has started to chip away at the conventions of televised broadcast cycles in favor of a new form of temporality—one governed by modes of attention, and the attendant terms and forms of connectivity among partisans across and among social media platforms. Some platforms swallow political objects like a black hole (e.g., Facebook) by disabling their users' ability to circulate to other platforms, while others (particularly YouTube, as discussed at length in Chapter Four) actively encourage spreading their objects (e.g., embedded videos) across a number of sites, platforms, and formats. Political communications can be said to be in a state of constant flux, in part because of the breakneck speed at which new platforms such as Twitter emerge on the media landscape, but also because of the degree to which the operating systems of such social media sites constantly change their underlying operational code. This is the language that governs the possibilities and limitations of the social Web. Political parties do not have the expertise, time, and money to keep a firm grip on this constantly mutating, innovating, and rules-shifting sphere of networked life. Yet, political parties and governments must be proactive in the networked environment. They've had to continuously intervene in the new 24/7-networked landscape to test out or otherwise prepare for possible elections and political crises. The result is a greatly expanded and partisan-filled networked landscape.

Throughout this book we draw attention to how such components of political networking both challenge and cultivate technological control over political life;

how such new actors, platform spaces, and objects contribute to the intensification of political partisanship, and the degree to which they both contribute to and undermine the ability to conduct a top-down hierarchical form of political management over the mainstream media, political parties, organs of government, and, ultimately, the voting public. Moving forward, all actors in a permanent campaign will inevitably struggle to keep pace with the new sites, times, and emergent partisan voices. Given the dizzying pace at which social media platforms update their back-end operational code, the task of researching such online-enabled political campaigning will also struggle to keep pace with technological change. Yet this is not simply a question of objectively understanding the techno-management of contemporary politics. Critical analyses of technologically mediated politics must also seek to develop a renewed set of concepts, including a political ethics, to frame our understanding of how networked computers, media, institutions and citizens invoke, enhance, or displace democratic life. While critics may decry the use of new media platforms by political war rooms and partisan operatives as merely networked forms of broadcasting that speak to voters without listening in return, permanent campaigning also raises serious ethical issues regarding the spheres of politics, the distinction between public and private, partisan publicity and voter/citizen privacy. Permanent campaigning is not merely a matter of expanding times and spaces, but rather of extending the opportunities and reach of partisanship across the entire political terrain.

At the outset of this book (Chapters Two and Three), we focus on the emergence of new political actors within networked platforms. In Chapter Two we set out to empirically discuss the emergence of partisan bloggers as new political actors. This chapter focuses on the contributions of bloggers to the mediated political landscape through an analysis of their shared blog posts. The chapter builds upon our previous research on blogging, research that demonstrated how shared links of blog posts can indicate the reach, popularity, and possible influence of a blogger—their status as a potential opinion leader both inside and outside of the political blogosphere. Here we offer two non-election case studies of blogging activity in Canada to better understand how bloggers relate to other political spheres, and specifically, how they invoke media stories, Wikipedia entries, and other online resources for their own political agendas. In short, this chapter qualifies the partisan role that new political actors play in various political campaigns and crises, paying close attention to the forms of media and communication that bloggers produce, circulate, and invoke during periods of heightened political activity.

Chapter Three conversely focuses on an altogether different form of actors: the publics that come into being on Facebook. Through an analysis of three case studies of online political activism on Facebook, Chapter Three conceptualizes the deployment of "issue publics" (Lippmann, 1993; Marres, 2005) on Facebook. We argue that the Facebook platform has developed a series of online spaces that bring together a public around specific political issues, mirroring in some respects the fragmented nature of networked political communications and campaigns. Using Maurizio Lazzarato's exploration of immaterial labor (2004), we demonstrate the need to further understand the networking of publics and their issues by considering how online platforms provide the material, communicational, and social means for a public to exist, and therefore define the parameters for assembling issues and publics and circumscribe a horizon of political agency.

Chapter Four in turn examines the rise of the network platform in online politics and provides an important historical context to this book's overall arguments, recalling the first Google-YouTube project developed to provide election information, during the 2007 national elections in Australia. Dubbed "Google Australia Votes," the project serves an as important case study that highlights a key shift in network architecture—one based upon the move from a portalized formatting of political information on the Web to a more socially mediated network that facilitates the mobilization of media objects—principally through the promotion of code that enables the embedding of video objects on and across multiple Web formats (blogs, news websites, Facebook, etc.). This chapter investigates how the Google project served to "mediatize" or provide a particular Web-based platform for Australia's political parties and online partisans. More broadly speaking, the Google project also served as a test for early Web 2.0 practices and conventions, in a newly emerging landscape that while managed by the Google portal, also facilitated an emerging network of distributed media objects.

Building upon the previous chapter's discussion of the networking and embedding of media objects on and across social media platforms, Chapter Five questions how the interfaces of social media platforms (e.g., blogs, Facebook and Twitter), particularly the vertical tickers that display stories and user-generated and shared posts, pose new challenges to political communications campaigns and scholarly research on networked dynamics of permanent campaigning. Vertically looped tickers highlight the increasingly fleeting nature of contemporary networked and socially mediated communications because they provide an intensely compressed space (interface) and time to have one's media objects—i.e., posts, replies, and comments—viewed by friends and followers. This chapter draws upon our research collaboration with the Canadian Broadcasting Corporation's news division in 2008 to understand how Canadian political parties strategically intervened in real time on Twitter during a broadcast political debate. The chapter argues that such a "live" form of politics is highly dependent upon keeping up to date with ever-shifting platform conventions and protocols, a slippery slope that also poses significant problems for research on online politics. While Chapter Five calls into question how political parties are struggling to keep up with the demands and exigencies of social media, its main goal is to highlight the always-on dimension of permanent campaigning, a space where politics can be conducted "live" at any point of the day or night.

The book's final chapter (Chapter Six) builds from the preceding case studies to suggest a new conceptual and methodological framework centered on the tracking of networked political objects to examine emerging forms of political campaigning on and across Web 2.0 platforms in the North American context. The proposed method seeks to identify the new strategies that make use of campaign texts, users, keywords, information networks, and software code to spread political communications and rally voters across distributed—and therefore seemingly unmanageable—spheres of online communication. The proposed method differentiates itself from previous Web 1.0 methods focused on mapping hyperlinked networks. In particular, we pay attention to the new materiality of Web 2.0 as constituted by shared objects that circulate across modular platforms. In this last chapter, an object-centered method is further developed through the concept of *traffic tags*—unique identifiers that enable the circulation of Web objects across platforms and consequently organize political activity online. By tracing the circulation of traffic tags, we suggest future research can map different sets of relationships among uploaded and shared Web objects (text, images, videos, etc.), political actors (online partisans, political institutions, bloggers, etc.), and Web-based platforms (social networking sites, search engines, political websites, blogs, etc.).

References

Alesina, A. & Roubini, N. (1992). "Political Cycles in OECD Economies." *The Review of Economic Studies* 59 (4), 663–688.

Callon, M., Lascoumes, P. & Barthe, Y. (2009). *Acting in an Uncertain World: An Essay on Technical Democracy.* Cambridge, MA: MIT Press.

Cario, E. (2010). "Pour se défendre, l'UMP achète 'Bettencourt.'" Available at http://www.ecrans.fr/Pour-se-defendre-l-UMP-achete,10347.html.

Chadwick, A. & Howard, P. (2008). *Routledge Handbook of Internet Politics.* New York: Routledge.

Cohen, J. E. (2008). *The Presidency in the Era of 24-Hour News.* Princeton, NJ: Princeton University Press.

Cushion, S. & Lewis, J. (2010). *The Rise of 24-Hour News Television: Global Perspectives.* New York: Peter Lang.

Dagnes, A. (2010). *Politics on Demand: The Effects of 24-Hour News on American Politics.* Santa Barbara, CA: Praeger.

Doherty, B. (2007). "Elections: The Politics of the Permanent Campaign: Presidential Travel and the Electoral College, 1977–2004." *Presidential Studies Quarterly 37* (4), 749–773.

Elmer, G. (2006). "Re-tooling the Network: Parsing the Links and Codes of the Web World." *Convergence* 12 (1), 9–19.

Esposito, R. (2005). *Catégories de l'impolitique.* Paris: Seuil.

Flanagan, T. (2009). *Harper's Team: Behind the Scenes in the Conservative Rise to Power.* Montreal: McGill-Queens University Press.

Foot, K. A. & Schneider, S. M. (2006). *Web Campaigning.* Cambridge, MA: MIT Press.

Gillespie, T. (2010). "The Politics of 'Platforms.'" *New Media & Society* 12 (3), 347–364.

Heclo, H. (1978). "Issue Networks and the Executive Establishment." In A. S. King (ed.), *The New American Political System* (pp. 87–124). Washington, DC: American Enterprise Institute for Public Policy Research.

Hillygus, D. S. & Shields, T. G. (2008). *The Persuadable Voter: Wedge Issues in Presidential Campaigns.* Princeton, NJ: Princeton University Press.

Howard, P. (2006). *New Media Campaigns and the Managed Citizen.* Cambridge, UK & New York: Cambridge University Press.

Jenkins, H. (2006). *Fans, Bloggers, and Gamers: Exploring Participatory Culture.* New York: New York University Press.

Klein, J. (2005). "The Perils of the Permanent Campaign." *Time.* Available at http://www.time.com/time/printout/0,8816,1124237,00.html.

Langlois, G., Elmer, G., McKelvey, F. & Devereaux, Z. (2009). "Networked Publics: The Double Articulation of Code and Politics on Facebook." *Canadian Journal of Communication* 34 (3), 415–433.

Lanham, R. A. (2007). *The Economics of Attention: Style and Substance in the Age of Information.* Chicago: University of Chicago Press.

Latour, B. (2005). "From Realpolitik to Dingpolitik or How to Make Things Public." In B. Latour & P. Weibel (eds.), *Making Things Public: Atmospheres of Democracy* (pp. 14–41). Cambridge, MA: MIT Press.

Lazzarato, M. (2004). *Les Révolutions du Capitalisme.* Paris: Les Empêcheurs de Penser en Rond.

———. (2006). "The Concepts of Life and the Living in the Societies of Control." In M. Fuglsang & B. M. Sørensen (eds.), *Deleuze and the Social* (pp. 171–190). Edinburgh: Edinburgh University Press.

Lovink, G. (2008). *Zero Comments: Blogging and Critical Internet Culture.* New York: Routledge.

MacCharles, T. (2010). "Stephen Harper's Former Campaign Director Paints Unflattering Picture of PM's Leadership." Available at http://www.thestar.com/news/canada/article/817693--stephen-harper-s-former-campaign-director-paints-unflattering-picture-of-pm-s-leadership.

Marres, N. (2005). "Issues Spark a Public into Being: A Key But Often Forgotten Point of the Lippmann-Dewey Debate." In B. Latour & P. Weibel (eds.), *Making Things Public: Atmospheres of Democracy* (pp. 208–217). Cambridge, MA: MIT Press.

Marvin, C. (1988). *When Old Technologies Were New: Thinking About Electric Communication in the Late Nineteenth Century*. New York: Oxford University Press.

Murray, S. K. & Howard, P. (2002). "Variation in White House Polling Operations: Carter to Clinton." *The Public Opinion Quarterly* 66 (4), 527–558.

Ornstein, N. & Mann, T. (2000). *The Permanent Campaign and Its Future*. Washington, DC: American Enterprise Institute Press.

Park, H. W. & Thelwall, M. (2003). "Hyperlink Analyses of the World Wide Web: A Review." *Journal of Computer-Mediated Communication* 8 (4), 66–81.

Schattschneider, E. E. (1960). *The Semi-Sovereign People: A Realist's View of Democracy in America*. New York: Holt, Rinehart & Winston.

Schmitt, C. (2007). *Theory of the Partisan: Intermediate Commentary on the Concept of the Political*. New York: Telos Press.

Scholz, T. (2006). "The Participatory Challenge." In J. Krysa (ed.), *Curating Immateriality: The Work of the Curator in the Age of Network Systems* (pp. 189–207). Brooklyn, NY: Autonomedia.

Zelizer, B. (1992). "CNN, the Gulf War, and Journalistic Practice." *Journal of Communication* 42 (1), 66–81.

CHAPTER 2

Political Blogging and Politics Through Platforms

Faceless and enigmatic, bloggers have polarized opinions about the political potential of online communication. While often accused of being crass, rumor-mongering "amateur monkeys," to use the words of "blogophobe" Andrew Keen (2007, p. 3), bloggers have also come to be known as remarkably efficient fact-checkers, prompting a number of high-profile resignations and embarrassments for political representatives (Drezner & Farrell, 2004). Seemingly never lacking an opinion on any matter political or partisan, the blogger has challenged both media and political spheres, in effect producing a new form of political communication that has subsequently been appropriated by journalists, politicians, and the public relations industry in general. Once reviled, mocked, and denounced by mainstream politicos, bloggers are now a well-established and respected class of political actors, and in some exceptional instances they have raised millions of U.S. dollars in a matter of hours for political causes (Gleicher, 2011).

Blogging gained prominence in the early 2000s through the development of user-friendly Web-based publishing platforms that greatly simplified and automated the more complex coding and design processes of Web authoring, thereby enabling users to focus their efforts on publishing and sharing information. The user-friendliness of blogs has been credited with producing a more dynamic and inclusive political sphere (Bruns, 2005), in stark contrast to the perceived hegemony of the pre-Internet mass media age, where a few media giants controlled the means of news reporting (McChesney, 2001). Political bloggers emerged in parallel with new types of political actors such as "citizen journalists," a term used for non–media-affiliated or amateur journalists who typically sought to publish on social and political events and issues (Gilmor, 2004). Where the traditional mass media system suffered from biases in reporting due to the profit-driven logic of editorial content (Hackett

& Zhao, 1998; McChesney, 2001), or because of the constraints of deadlines and overreliance on official sources (Tuchman, 1978), blogging exemplifies a radically different dynamic that allows citizens not only to produce content but also to publicly assess journalistic contributions (Bruns, 2005).

Most research on political blogging, however, has largely abstracted itself from questions of software platforms in favor of questions about social and political effect; that is, how bloggers pose challenges to political roles, discourses, and debates. This research has mainly focused on characterizing bloggers as a new political voice rather than interrogating the networked software structures within which such voices are framed. Such an oversight is problematic given the rise of social networking platforms and other online political applications (e.g., online petitions, meetup.com) that bloggers increasingly react to and use to spur political action, rationalize their partisan position, or frame their participation in political discourse. It is this technologically mediated *partisan* blogosphere that captures our present attention and frames this chapter's analysis of bloggers' interventions in political crises.

Canada is perhaps a ready-made jurisdiction for the analysis of *partisan blogging*: the act of publishing, networking, and at times, organizing in support of a political party. Canada's political parties do not typically organize their own blogrolls. Instead, Internet supporters of each of Canada's main political parties group together to form partisan blogrolls (Elmer et al., 2009), which in their simplest form provide hyperlinked lists of individual blogs sharing a common set of political banners. Liblogs (the blogroll of supporters of Canada's Liberal Party), for example, emerged after the party's defeat by the Conservatives in January 2006. The growth of the Canadian blogosphere can be explained—at least in part—by the ongoing leadership races (and crises) within the Liberal Party, and subsequent attempts by other partisan bloggers (Tory bloggers, New Democratic Party bloggers, Green bloggers, etc.) and self-organized nonpartisan political bloggers[1] to antagonize, criticize, and otherwise comment on the demise of the once-dominant Liberal Party.

While Canadian political blogrolls were spurred by the crisis that the Liberal Party faced from 2006, right from the beginning, they insisted on their independence from the official apparatuses of the country's established political parties. Perhaps seeking to cultivate their own status as political actors in a field long dominated by political insiders, journalists, and political pundits in the mainstream media, bloggers reveled in an in-between space, one that offered room for both dissent and party loyalty. However, as bloggers' numbers increased and debates flourished, evidence of links to the political establishments mounted.

The political advisors around Conservative Prime Minister Stephen Harper, for example, discussed the need to monitor the blogosphere and use partisan bloggers to "get out stories that were not quite ready for the mainstream media" (Flanagan, 2009, p. 232). The Liberal and New Democratic Parties have similarly sought to maintain relationships and open communications with select groups of partisan bloggers in the hopes that they will contribute to the goals of central party offices. Although bloggers are feared and mistrusted by the political parties, background campaign and communications staff have used them as political proxies: as voices that serve to heighten, reshape, or distract from particular political debates.

In Chapter One we conceptualized partisan political bloggers as "mobile actors," individuals who work semi-independently on the peripheries of official party networks, spaces, and temporalities. Mobile partisans, however, also enter into the mainstream from time to time, or in some cases entirely dominate and define a political event or debate. In this chapter we seek to better understand the contours of such mobility, the entrance and exit routes of new political actors into heightened periods of political crisis. As *mobile* partisans, bloggers move not only in between and among pre-existing political discourses and professions, but also within and across media landscapes. Anecdotally, then, we know that bloggers are sometimes (particularly during election campaigns) courted and consulted by political parties and are encouraged to play a part within a more rigidly controlled partisan campaign. Here we look beyond such anecdotal, insider banter—typically filled with bloggers' ego-filled claims to political influence—to understand the political *actions* of bloggers in particular, to understand their relationship to parties, the mainstream media, and new user-generated platforms in the production of a networked form of politics. The studies presented in this chapter seek to determine the specific actions and conventions of partisan blogging at different times of the political calendar to determine *when, how, and if* they intervene in ongoing partisan campaigns and communications. We are less concerned with the political discourse and debates of bloggers; that is, the topics, rhetorics, and language of these social media actors (conversely, see Chapter Three for a discussion on issue-publics on Facebook). Rather, through an analysis of the platforms, documents, and digital objects (blog posts, online videos, newspaper stories) that bloggers link to in their posts during particular political periods, we can begin to understand the contours of bloggers' partisan contributions: how bloggers' links to documents, platforms, tools, and applications suggest different forms of political activity, ranging from media commentary to fund-raising, historical research, networking, and the organization of protests and other events.

Partisan mobilization, however, is not a stable factor. In a permanent campaign, partisan mobilization reaffirms the imperative of constant readiness to action, even at a moment's notice. The studies compiled in this chapter analyze mobile partisanship in different time frames of the permanent campaign in an attempt to quantify and qualify heightened periods of partisan activity. We begin by providing an overall picture of blogging activity by charting the daily production of blog posts in the Canadian partisan blogosphere. Next, we focus on two recent political crises in Canada: the coalition crisis (November 18, 2008–December 9, 2009) and the prorogation crisis (December 28, 2009–January 31, 2010). Both events took place under a Conservative minority government in a Westminster-style Parliament that requires a majority vote from its members for the passage of legislation. Without a majority of votes, the governing Conservative Party was always wary of being outvoted in the House of Commons and, in effect, removed from office, over any number of political issues, proposed laws, or legislative maneuvers.

In late November 2008, the Conservative Party, under the leadership of sitting prime minister Stephen Harper, precipitated one of Canada's most complex and compelling political dramas in recent memory. After the government tabled legislation that would have gutted the opposition parties' fund-raising apparatuses, on December 1, 2008 the Liberal and New Democratic Parties joined forces to propose a coalition government with the support—but not explicit participation—of the separatist Bloc Québécois. However, before the opposition parties were able to officially defeat the government with a binding vote in the House of Commons, the prime minister "prorogued" Parliament, a maneuver typically used to bring the normal timeline of a parliamentary session to an end. The opposition parties accused the government of abusing the typically perfunctory maneuver, and on December 4, 2008, suggested that Canada's official—though largely ceremonial—head of state, Governor General Michaëlle Jean, decline the prime minister's request to close Parliament because he had lost the confidence of the majority of the House of Commons. Needless to say, the announcement of a coalition alternative, Harper's prorogation request, and the governor general's reluctant acceptance of the parliamentary closure produced an intense period of partisan bickering that occurred both online and offline. The crisis, and the subsequent period of partisan campaigning that lit up the Internet, came to an end after the Conservative government unveiled a revised budget plan that met most of the demands of the Liberal Party (which itself remained mired in an ongoing leadership crisis).

The second case study focuses on blogger activity during another instance of Prime Minister Stephen Harper seemingly preempting the everyday business of Parliament for partisan reasons. This so-called "prorogation crisis" (December 28, 2009–January 31, 2010) was spurred by the decision by Harper to suspend Parliament once again on December 30, 2009 until the beginning of March 2010. Like the preceding coalition crisis, the decision was seen by opposition parties as another cynical ploy by the Conservative government, on this occasion to avoid further investigations into the alleged torture of Afghan detainees by Canadian forces. As with the coalition crisis, the Internet witnessed substantial activity in the weeks following Harper's move to suspend the House of Commons. Together, the coalition and prorogation crises are commonly viewed as two of the very few political miscalculations from a prime minister widely lauded for his political skills.[2]

Expressions of Mobilization: The Challenges of Hyperlink Research

While we, in our work at the Infoscape Centre for the Study of Social Media, have previously conducted research using content analysis of political blog posts (Elmer et al., 2006), in particular during a Liberal Party leadership campaign to measure opinions of candidates, such an approach quickly revealed itself to be too time-consuming given the exponential growth of blog posts during heightened periods of partisan politics. Furthermore, we also realized that treating all blogs as equal is not representative of the communicational dynamics at work in the political blogosphere. Simply put, some blogs are read, commented upon, and referred to more than others. Our approach at the time treated each blog post as an individual unit of analysis; however, a fundamental characteristic of political blogs is that they offer numerous possibilities for connectivity. In addition to commenting on political matters, bloggers' posts often refer to sources, opinions, and analyses available on other platforms in order to push their point forward. There are important communicational dynamics at stake in the political blogosphere, stemming from a networked performativity or connectivity. Bloggers not only offer a unique form of communication (embodied in the blog post), they also consistently network with other bloggers, political and media actors, and platforms to gain an audience and acquire and enhance their political capital. Our approach thus seeks to measure the importance of specific issues and actors in the blogosphere at different times of the permanent campaign. As such, our research focuses on measuring levels of visibility (or, conversely, invisibility) of specific Web objects (i.e., blog posts, news articles) that frame the discourse of bloggers.

The most reliable metric for providing such information is the hyperlink, particularly those attached to specific blog posts. When a Web object such as a blog post or news article is linked to in the blogosphere, it typically indicates that the specific object is worthy of interest and conversation. Other metrics can be used, such as the ranking on Google or other popular search engines, although the logic of this ranking is a well-kept proprietorial secret, depriving the researcher from crafting their own sample. Another possible metric involves collecting and conducting a content analysis of comments on specific blog posts; however, bloggers have the power to disallow, delete, or delay the posting of comments from readers, thereby producing an unreliable sample. Overall, the common use of blog post hyperlinking makes it a reliable indicator of visibility on the blogosphere. Among existing participatory media platforms, the prominence of hyperlinking on blogs is noticeable, as other platforms create connections not only through hyperlinks but also through embedding text and videos. The one limitation of the hyperlink-based method is the embedding of videos, as these Web objects cannot be identified through the collection of hyperlinks on blog posts. As such, our analysis of the relationship between bloggers and other networked spheres does not include YouTube (but see Chapter Four for an analysis of political campaign videos).

Hyperlinks have been an object of study and a metric for highlighting communicational dynamics since the beginning of Web studies. With regards to research on political issues, the hyperlink was considered first as a reliable metric to examine the formation of political networks, both online and offline. Such was the approach of Garrido and Halavais (2003) when examining the hyperlink network around the Zapatista movement in order to analyze how a local struggle gained global awareness and was linked to other antiglobalization issues. Another notable method of hyperlink analysis in online political communication focuses on creating maps of hyperlinks to show political connectivity (Park & Thelwall, 2008). The issue-network approach (Marres & Rogers, 2005) to hyperlink analysis, which entails examining the hyperlink networks surrounding a selection of websites involved in a particular social or political issue, has been particularly influential on our methods. Such an approach makes use of automated Web crawling and hyperlink visualization tools (e.g., issuecrawler.org) to understand the online relationships among political actors over time.

The arrival of blogs has complicated hyperlink research because of the dramatic increase in the use of hyperlinks. In the Web 1.0 environment, a hyperlink was a straightforward deictic pointer that suggested a relationship, whether positive or negative, between two Web pages. All hyperlinks were judged equally. The blog format, in contrast, introduces new categories of hyperlinks. Examples include: the fixed hyperlinks that are constantly featured on the sides of a Web page (i.e., blogroll links, links to "blogs I read," recommended blogs, favorite websites, etc.); hyperlinks from online advertisers (e.g., Google AdSense) designed to generate revenue; and the hyperlinks embedded in blog posts as part of a discussion on current issues. These hyperlinks signify altogether different spheres and relationships, such as political affiliation and revenue generation. Apart from using the blogrolls' lists of links to assemble the blog samples, we put these links aside.

As we detail in this chapter, our analysis of partisan blogging focused on collecting blog post hyperlinks from the main partisan blogrolls in the Canadian English-speaking blogosphere: the Blogging Dippers (New Democratic Party, or left of center), the Blogging Tories (Conservatives, or right of center), Green bloggers, the Liblogs (Liberals, commonly viewed as a centrist party), and political bloggers listed with the nonpartisan blogroll. Full RSS feeds from individual blogs were used to collect and store blog posts daily through the Gregarius data collection tool. Our previous research on blogs entailed a similar data collection process, followed by use of customized software tools and scripts that identified the total number of posts in our sample as a daily updated graph (the blogometer tool) and collected hyperlinks embedded in blog posts (the blog link ripper). The blog link ripper enabled us to unearth the most-linked-to Web objects at a given period (i.e., most-linked-to blog posts, most-linked-to news stories) in order to determine the most popular and widely discussed issues and perspectives in the blogosphere.[3]

These software tools are useful when undertaking a longitudinal and comparative analysis of the Canadian blogosphere between August 2008 and July 2010—a period marked by a series of federal and provincial elections, as well as the two political crises that constitute the present case studies. This is true of the blogometer in particular. By measuring posting activity, the blogometer charts moments when the blogosphere is particularly active (defined as an increase in blog posts). The resulting chart (Figure 1) offers a simple weekly analysis of blog posts by all partisan bloggers in our sample between August 2008 and July 2010 (comprising 969 bloggers). Rising activity during the week and falling activity during the weekend express a kind of sinusoid wave that forms the baseline of blogging activity. A closer day-to-day examination of blog posts shows a similar rise and fall of activity, with activity rising in the morning, followed by a sustained period of stronger activity during the day and then a slow trailing off at night. Bloggers are most active Mondays through Fridays, while the weekends mark a low point of activity. Indeed, Friday, Saturday, and Sunday nights hover around three posts per hour for our entire sample.[4] Such a routine wave pattern is not limited to Canadian blogs; indeed, research into the Australian blogosphere (Bruns et al., 2010) reveals a similar "heartbeat" rhythm for weekly blogging activity that corresponds to higher blogging rates during work hours.

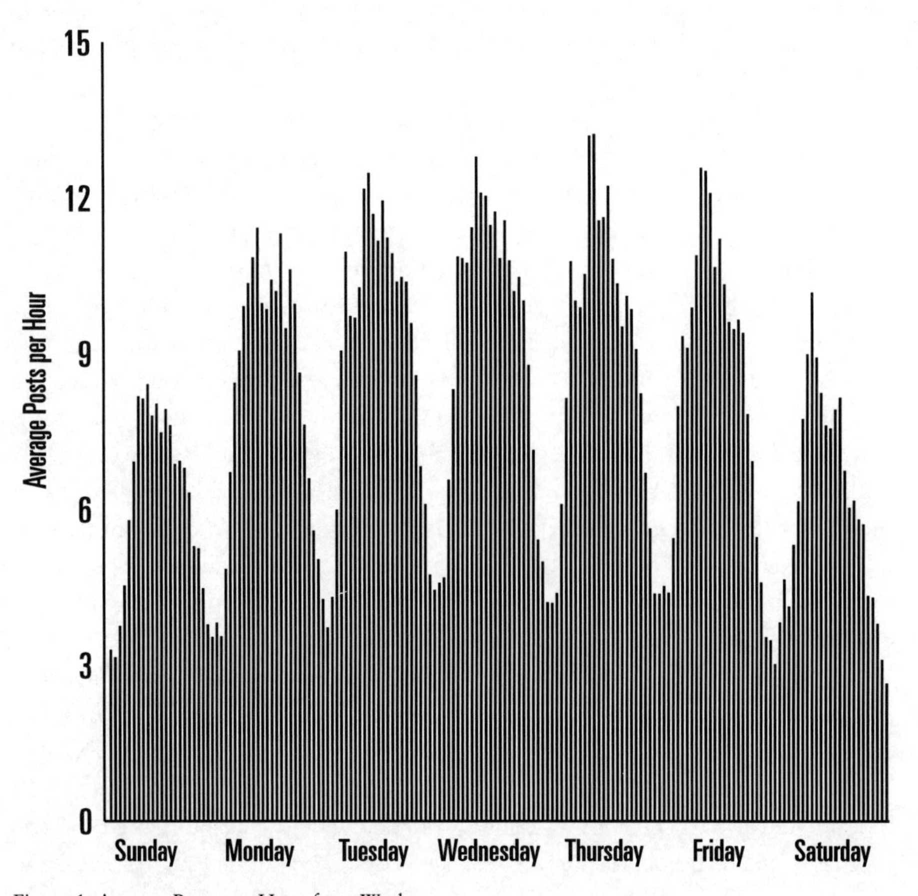

Figure 1. Average Posts per Hour for a Week.

Figure 1 suggests that political activity might follow predictable or at least common patterns for most bloggers. Deviations from such a routine—such as daily spikes in activity or increased blogging over the weekend—might subsequently indicate the emergence or full-blown development of a political crisis. Figure 2 provides a clear indication of heightened blog activity during the two political crises under study.

From August 28, 2008 to April 9, 2009—the date when we updated our blog sample (adding new blogs and removing inactive or deleted blogs)—the blogosphere averaged 229.52 posts per day. This represents a significantly higher amount than the average 147.52 posts per day gauged from April 28, 2009 to August 18, 2010. Conversely, as Figure 2 demonstrates, the coalition crisis in the fall of 2008 corresponds to the most active period of blogging activity in the partisan blogosphere during the period August 2008–July 2010.[5]

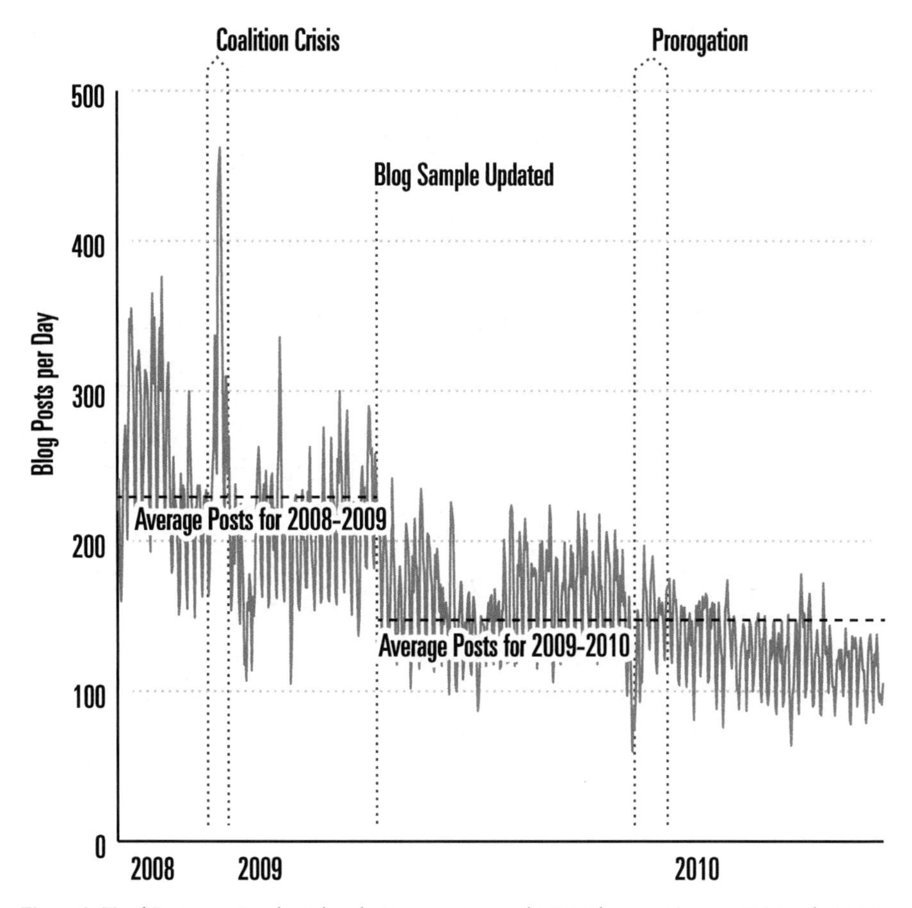

Figure 2. Total Posts per Day by Liberal, Conservative, and NDP Bloggers (August 2008–July 2010).

As can be seen in Figure 2, the coalition crisis corresponded with greater-than-average posting activity in the Canadian blogosphere, thus indicating a high level of blog posting—presumably partisan attacks against, or defenses of, the governing Conservative Party. By contrast, the prorogation crisis (see top right of

Figure 2) did not provoke a similar surge of activity in the Canadian blogosphere. Merely charting the productivity of bloggers obviously cannot explain the difference in blogging activity and practices during these two crises. Rather, to better understand the role of bloggers during these two political crises we turn to a more qualitative analysis of hyperlinks deployed and embedded by bloggers in an effort to understand not just the quantity of blog posts but also the networked characteristics of such posts. That is, what bloggers were referring to (linking to) and invoking in their responses to the political crisis. These two periods, then, offer a chance to study the relationship of the blogosphere to emerging social media platforms, other political actors, and the mainstream media during heightened periods of political activity.

In this study, we focus on the relationship of blogs to different Web platforms and domains, and not to specific URLs. A domain is a unique address for a website on the Internet; for example, the URL http://www.infoscapelab.ca/node has the domain www.infoscapelab.ca. Everything after ".ca" specifies a location *within* that domain. Moreover, domain names provide us with a broad yet helpful form of coding different spheres on the Web: for instance, .edu, .org, .gov, and .ca indicate specific national or occupational sectors (education, not-for-profit organization, U.S. government, and Canadian, respectively). However, domains only indicate very broad categories, and are sometimes too vague to be of significance. For example, domain names ending with ".com" can be applied to businesses, mainstream media, alternative media, and personal websites, among many others. Similarly, the domain-name system has become even more diffuse with the introduction of Web 2.0 and social media platforms, many of which use the generic .com domain. Our approach consequently seeks to identify the type of websites linked to in blog posts—whether they exist to inform, entertain, provide knowledge, connect people, or showcase user-generated content—in order to identify the broader communicative spheres (i.e., news and analysis, user-generated content, etc.) that bloggers are engaging with. The study of hyperlinks can provide insight into the types of sources, platforms, objects, and texts used by different kinds of bloggers (in this case, from the government, the mainstream media, alternative media, political parties, and nonpartisans). The practice of connecting to other spheres via hyperlinks reveals how bloggers situate a particular issue by framing it (which media sources are being used, for example) and connecting it to families of actors (e.g., professional politicians through linking to party websites, and the more general public through linking to social networking platforms such as Facebook). Bloggers' encoded hyperlinks from the coalition and prorogation crises thus offer the basis for quantifying and qualifying the performative and connective aspects of mobile partisanship, examining the practice of structuring blog participation at specific moments through processes of affiliation and mobilization of specific communicative spheres.

In total, 15,867 hyperlinks from the coalition crisis period and 13,235 hyperlinks from the prorogation crisis were collected and coded, first by identifying their different Web spheres and more specific subcategories. For instance, a link to the www.globeandmail.com domain (the website for Canada's leading national newspaper) is coded as belonging to the "News and Analysis" Web sphere, then further coded under the subcategory "Mainstream Media." The News and Analysis sphere also includes subcategories such as alternative media, online magazines, etc. The "Web sphere" concept is taken from Kirsten Foot's analysis of online political campaigns. Foot's conception of Web spheres seeks to recognize the intersection between communication and organizing on the Web through links and other online indicators. Drawing on Taylor and van Every (2000), Foot (2006) defines the Web sphere as "a collection of dynamically defined digital resources spanning multiple Web sites deemed relevant or related to a central theme or object" (p. 89).

Web Spheres	Sub-categories of Web Spheres							
Blogs	Independent Political Blog	Professional Blog	Independent Blog	Professional Media Blog				
For-Profit	Corporate Website	Entertainment Website	Gaming Website					
News and Analysis	Alternative Media	Independent Media	Mainstream Media	News Aggregator	Online Magazine			
Professional Politics	Political Party Website	Political Website	Politician's Website					
Issues and Advocacy	Community Group	Non-Governmental Organizations	Religious Website	Union Website				
Knowledge Resources	Documentary Website	Education Website	Library Website	Online Encyclopaedia	Online Resource	Petition Website	Polling Website	Software Development Website
Indexing	Blog Search Engine	Blogroll	Index	Portal	Search Engine			
User-Generated	Blog Aggregator	Blog Platform	Forum	Participatory Media	Personal Website	Social Network Site		
Governmental	Government Website	International Governmental Organization						

Figure 3. Web Spheres Coding Sheet for Hyperlinks.

The Coalition Crisis: Platform Connectivity

The coalition crisis began with a promise from the newly elected Conservative government to respond to a looming economic recession with the release of an economic update, a mini-budget of sorts. As part of the mini-budget, the government announced a plan to cut government subsidies to political parties.[6] Fearing for their own survival, the opposition parties signed a historical pact to form a coalition government. The government subsequently prorogued—or shut down—Parliament to prevent a vote of no confidence (the formal vote in the House of Commons needed to officially defeat a sitting government). Consequently, as noted in Figure 1, activity in the blogosphere reached unprecedented levels. Usually, Liberals bloggers averaged 68.46 posts per day, NDP bloggers 28.38 posts per day, and Conservatives bloggers 132.69 posts per day, from 28 August 2008 to 9 April 2009. During the coalition crisis, these numbers jumped considerably: NDP blog activity increased 30.59 percent to 37.06 posts per day; Liberal blogs jumped 22.17 percent to 83.64 posts per day; and Conservative blogs rose 18.09 percent to 156.59 posts per day.

Figure 4 provides a summary of the Web spheres that partisan bloggers supporting Canada's three federal parties (Liberal, Conservative, and NDP), in addition to Green Party and nonpartisan bloggers, linked to during the coalition crisis period. Each cell includes the percentage of the blogroll's total links.

	Conservatives	Greens	Liberals	NDP	Nonpartisan
News / Analysis	50.14%	24.19%	46.75%	53.20%	38.19%
Blogs	20.33%	29.61%	19.03%	15.39%	13.19%
User-Generated Content	10.12%	14.88%	6.60%	4.93%	10.16%
Knowledge Resources	4.19%	4.96%	5.23%	6.90%	3.02%
Professional Politics	3.49%	4.81%	8.42%	3.45%	2.47%
For-Profit	3.88%	9.15%	3.58%	3.76%	3.02%
Issues & Advocacy	2.95%	7.75%	2.48%	3.57%	23.35%
Indexing	2.81%	1.55%	4.51%	3.63%	4.40%
Governmental	2.11%	3.10%	3.41%	5.17%	2.20%

Figure 4. Percentages of Links to Web Spheres During the 2008 Coalition Crisis.

Overall, News and Analysis links, usually to mainstream media sources, dominate linking practices—with the NDP (53 percent), Liberal (47 percent), and Conservative partisan blogrolls (50 percent) linking to this sphere. The Greens and nonpartisan bloggers, however, were by far less inclined to invoke media reports during the crisis. Only 38 percent of links from nonpartisan bloggers were directed at media platforms and sites, with less than 25 percent for Green bloggers. Internal links within the blogosphere (bloggers linking to other blog posts) was the second most common form of hyperlinking, except for the nonpartisan bloggers, who instead linked to the "Issues and Advocacy" sphere. After News and Analysis and internal blog links, connections to other Web spheres falls off dramatically. The third-most-linked-to sphere for the Greens (15 percent), Conservatives (10 percent), and nonpartisans (10 percent) was the "User-Generated Content" sphere (typically social media platforms), while Liberal bloggers (8 percent) more commonly linked to the sphere of "Professional Politics," and NDP bloggers (7 percent) chose to link to the "Knowledge Resources" sphere.

An analysis of hyperlinking patterns over the course of the coalition crisis (see Figure 5) further qualifies the linking practices of political bloggers over the 22-day period of the crisis. Figure 5 confirms the predominance of the News and Analysis sphere, which at no time during the crisis falls behind any other sphere of websites. There is a significant rise in links to the News and Analysis sphere on December 1, 2008,[7] from just over 150 links to almost 350 links. Links to User-Generated Content also rose sharply on December 1, 2008, with many bloggers linking to Facebook groups that formed in support or opposition to the proposed coalition government. There was also a significant spike in blog links to Professional Politics on the same date.

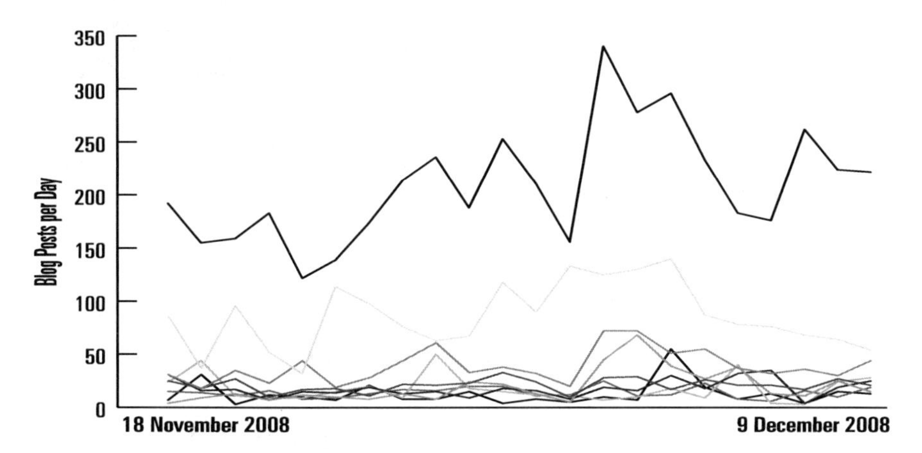

Figure 5. Total Posts per Day by Type of Domain.

A closer analysis of the News and Analysis sphere (Figure 6) shows that bloggers linked most frequently to mainstream media sources, and typically to major Canadian news outlets. Theglobeandmail.com (Canada's national newspaper of record), cbc.ca (the national public TV and radio broadcaster), canada.com (an important newspaper conglomerate), and ctv.ca (one of Canada's major national private broadcasters) lead in overall linking: the CBC received 52 links; *The Globe & Mail* received 45 links; and CTV trailed with 27 links. Links referred to news articles that discussed the upcoming Conservative budget or details of the coalition talks. When breaking down linking practices by blogrolls, we find that Liberal and NDP bloggers both linked most to the CBC and *The Globe & Mail*. By contrast, Conservative bloggers linked more to the national television outlet CTV (17 links) and the conservative daily newspaper the *National Post* (10 links), as well as to international news media such as *The Guardian* (11 links) and the BBC (11 links). These links suggest some ideological framing at work. Conservatives link to outlets seen as right of center (the *National Post* and CTV), whereas NDP and Liberals link to centrist outlets (*The Globe & Mail* and, to a lesser degree, the CBC).

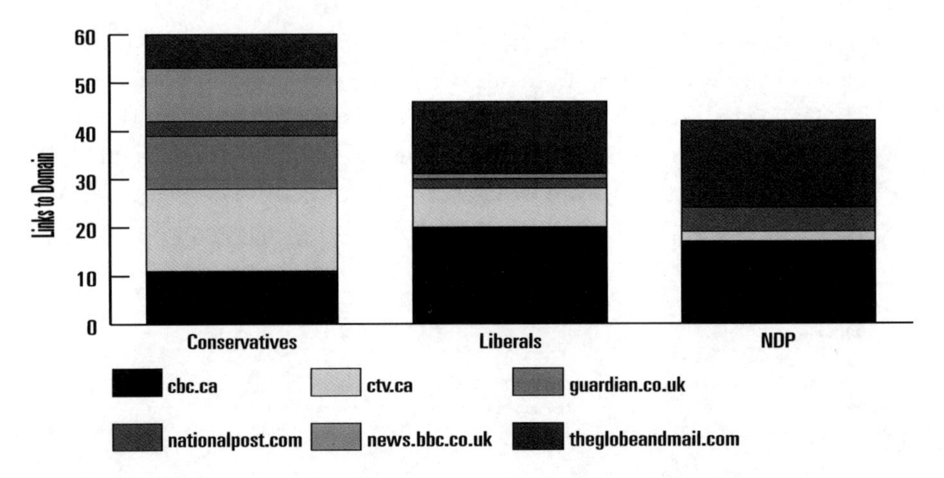

Figure 6. Top Domains in the "News and Analysis" Sphere by Party for December 1, 2008.

Overall, the predominance of mainstream media in the hyperlink samples requires us to rethink the notion of bloggers as citizen journalists. The reliance on mainstream news sources rather than independent news sources shows that bloggers are not generally interested in perspectives that are ignored or not covered by the mainstream media. As such, it could be said that while bloggers might offer a different perspective on issues of common interest, they are interested in linking these perspectives with mainstream news discourses. The analysis of links does not show whether bloggers agree or disagree with what is portrayed in the news (this would require a content analysis), but it reveals that the type of journalistic practices developed by bloggers is framed within mainstream discourses.

Closer inspection of the linking of each blogroll on December 1, 2008—the day when news about a potential coalition government appeared—also reveals a rise in links to the Professional Politics sphere. This distinction is particularly pronounced for the Liberal and Conservative partisan bloggers who, as seen in Figure 7, link to both their own party and rival campaign websites. Liberal blogs linked to the Victory Fund (8 links), a donation page for the Liberal party, as well as the Liberal Party's website (5 links). The Conservative blogs, on the other hand, linked to the Rally for Canada website (7 links) launched by Conservative blogger Stephen Taylor "to support the grassroots efforts already underway to organize rallies in cities all across this great country" and to gather together "people who want Parliament to stop bickering and get back to work."[8] Liberal Member of Parliament Michael Ignatieff, whose name had circulated as a possible replacement for the failing Stéphane Dion as head of the Liberal Party at that time, also received links from the Liberals and the NDP (3 links).

These findings demonstrate mobile partisanship at work: at times of crisis, bloggers mobilize in support of their party's stance on a given issue. The bloggers are not simply framing their discourse; they are inviting their readership to take action and further promote the cause. This type of transition from publishing partisan messages to inviting readers to take action reveals the rise of a process of motorization of bloggers. By motorization, we mean the ensemble of practices through which a machine is built not only to spread a message across Web spheres, but also to enhance the performativity of this message. Bloggers, in this case, not only relay a message from the party website, they also embed this message to a campaign website and thus further publicize a political strategy.

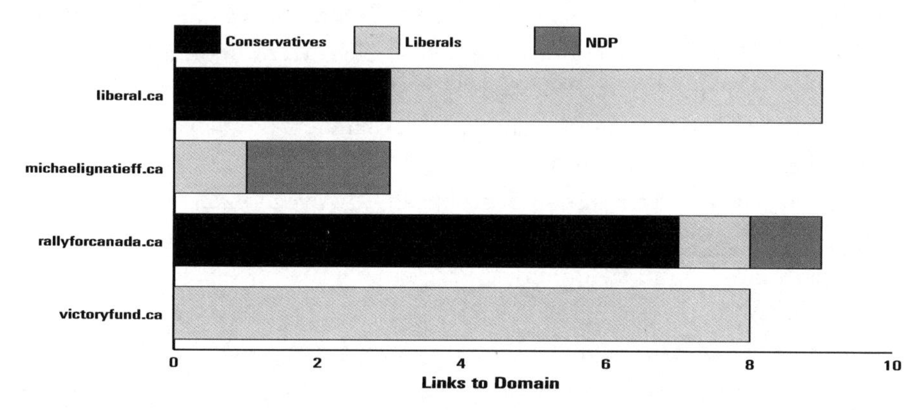

Figure 7. Links to Top Professional Politics Domains by Party for December 1, 2008.

The Prorogation Crisis: From Blogging to Social Networking

Almost one year after the coalition crisis, the Canadian minority Parliament edged toward dissolution again. Questions regarding the Canadian Forces' handling of detainees during the war in Afghanistan as well as concern over the economy intensified political antagonisms, giving the ruling Conservatives little room to maneuver. Prime Minister Harper once again prorogued Parliament, shutting down the House of Commons on December 30, 2009. Harper argued that the government needed a chance to reconnect with the electorate before introducing a budget in the spring. Dimitri Soudas, press secretary for the prime minister, explained that the government wanted to "give Canadians an overview of where [they] will be taking the country over the next little while."[9] Parliament was prorogued until March 3, 2009, a date chosen because it fell after the end of the 2010 Winter Olympic Games held in Vancouver, British Columbia.

This second parliamentary prorogation caused an uproar in the political sphere and on social networking platforms,[10] and it was heavily debated in the mainstream media. While opposition parties saw another prorogation as yet more proof that Prime Minister Harper was "despotic" and "arrogant,"[11] the most noticeable characteristic of the crisis was that it marked the rise of Facebook as a central organizing platform for political action, both online (through the promotion of petitions) and offline (through the organization of protests and rallies). Activity on Facebook intensified as opponents of prorogation organized rallies nationwide to express their concern. While Facebook had witnessed some increased activity during the coalition crisis, with a series of pro- and anticoalition protests organized via Facebook groups, public reliance on Facebook became even more pronounced during the prorogation crisis. In particular, the "Canadians Against Proroguing Parliament" Facebook group became one of the most important groups in recent Canadian history, with upwards of 40,000 members within days of its creation and reaching more than 200,000 members by the end of January 2010. The "Canadians Against Proroguing Parliament" group also reached beyond entrenched partisans and politically involved online citizens to a new public. Most members, according to a survey conducted by the Rideau Institute, had never joined a political Facebook group before.[12]

The period immediately following multicity rallies against the prorogation (December 29, 2009 to January 23, 2010) is likewise important to understanding how bloggers managed to sustain the various campaigns on and across the Internet's many spheres. Unlike during the coalition crisis, however, blog activity during this period did not deviate much from the average. The Liberals averaged 50.64 posts per day, the NDP averaged 23.78 posts per day, and the Conservatives averaged 73.10 posts per day from April 28, 2009 to August 18, 2010 (see Figure 8). Average posts per day during the prorogation crisis stayed relatively flat (+3.0 percent for the Conservatives and +3.08 percent for the Liberals) or even dropped (-15.08 percent in the case of the New Democrats).

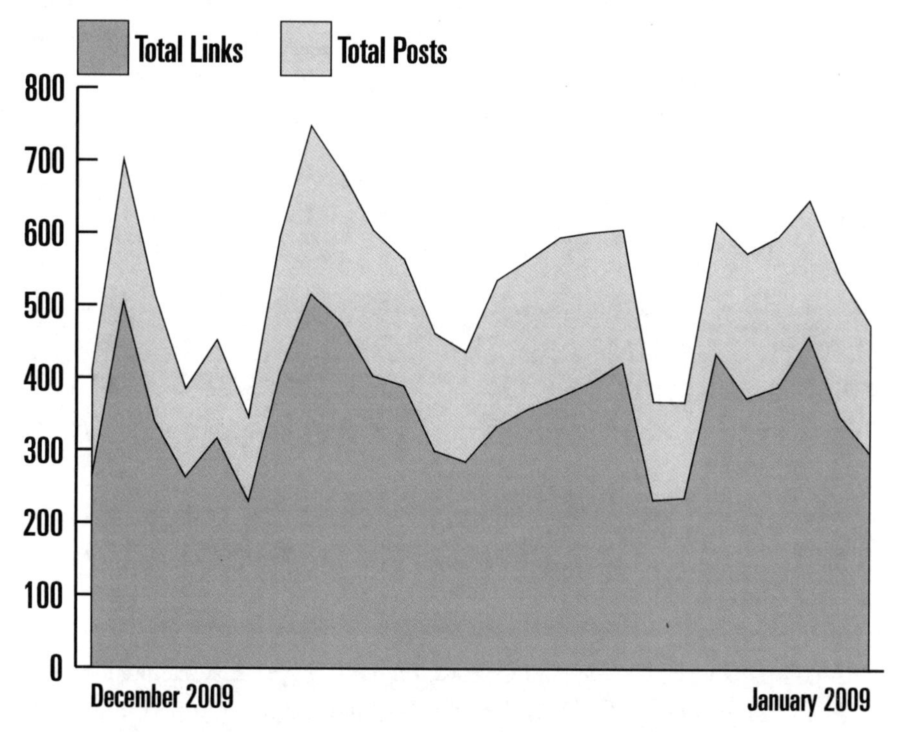

Figure 8. Activity in the Blogosphere During the Prorogation Crisis: Posts and Hyperlinks.

Activity in the blogosphere during the prorogation crisis remained fairly consistent at around 150 to 200 posts per day. The encoding of links in blog posts, however, spiked occasionally. These spikes in linking practices, as depicted in Figure 8 above, correspond to key moments in the news cycle. The spike on December 30, 2009 coincides with the announcement of the prorogation of Parliament; the spike on January 5, 2010 occurred after a few news articles discussed online opposition to the prorogation; and the spike on January 21, 2010 resulted from the release of a major poll about Canadian citizens opposing prorogation. Bloggers played a much more peripheral role during the prorogation crisis, while the epicenter of political activity online shifted to other spheres and platforms, such as user-generated content and especially social networking sites such as Facebook.

Returning to bloggers' links during the prorogation crisis (see Figure 9), we clearly see a shift in activity from commenting and framing the crisis through news reports to connections made within the User-Generated Content sphere. Yet, as during the preceding coalition crisis, the News and Analysis sphere received the largest percentage of links from all blogrolls (albeit a substantially reduced percentage): Conservatives (46.50 percent), Liberals (38.63 percent), and the NDP (36.47 percent). The nonpartisan blogroll linked considerably less to this sphere (24.37 percent) than did the other blogrolls.

	Conservatives	Nonpartisan	Liberals	NDP	Greens
News & Analysis	46.50%	24.37%	38.63%	36.47%	37.84%
User-Generated Content	11.86%	26.06%	21.82%	22.00%	17.57%
Blogs	20.65%	22.33%	15.08%	14.98%	22.97%
For-Profit	5.88%	10.49%	5.19%	5.31%	6.76%
Knowledge Resources	5.48%	4.30%	5.96%	5.05%	1.35%
Governmental	2.25%	3.61%	5.92%	5.65%	6.76%
Issues & Advocacy	3.63%	3.54%	2.25%	5.65%	2.70%
Indexing	2.41%	4.57%	2.16%	3.42%	0.00%
Professional Politics	1.35%	0.73%	2.98%	1.46%	4.05%

Figure 9. Hyperlinks to Web Spheres During the 2009 Prorogation Crisis.

In comparison with the coalition crisis, Figure 9 reveals a substantial increase in links from bloggers to the User-Generated Content sphere during the prorogation crisis from the following: the nonpartisan blogroll (26.06 percent), the Liberal blogroll (21.82 percent), and the NDP blogroll (22 percent). During the coalition crisis, the percentage of links to the User-Generated Content sphere was below 10 percent for the Liberals, the NDP, and the Nonpartisan (6.60 percent, 4.93 percent and 10.16 percent, respectively). The Blogging Tories linked to the User-Generated Content sphere less than any other blogroll (11.86 percent), an indication that Conservative partisans were not as keen to organize and frame the prorogation as an event worthy of networked action. Conservatives were more likely to link to others in the political blogosphere, indicating that they were mostly debating with and commenting upon other bloggers' posts. And unlike the Liberal (15.08 percent) and New Democrat bloggers (14.98 percent), nonpartisan bloggers also contributed a sizeable percentage of their hyperlinks to other blogs (22 percent).

If we further break down and analyze links to the User-Generated Content sphere (including Wikipedia for comparative purposes) we see in Figure 10 that the so-called "ABC" movement ("Anyone but Conservative"; i.e., the Liberals, NDP and Greens) gravitated to Facebook and Twitter, while Conservative bloggers preferred Wikipedia and YouTube.

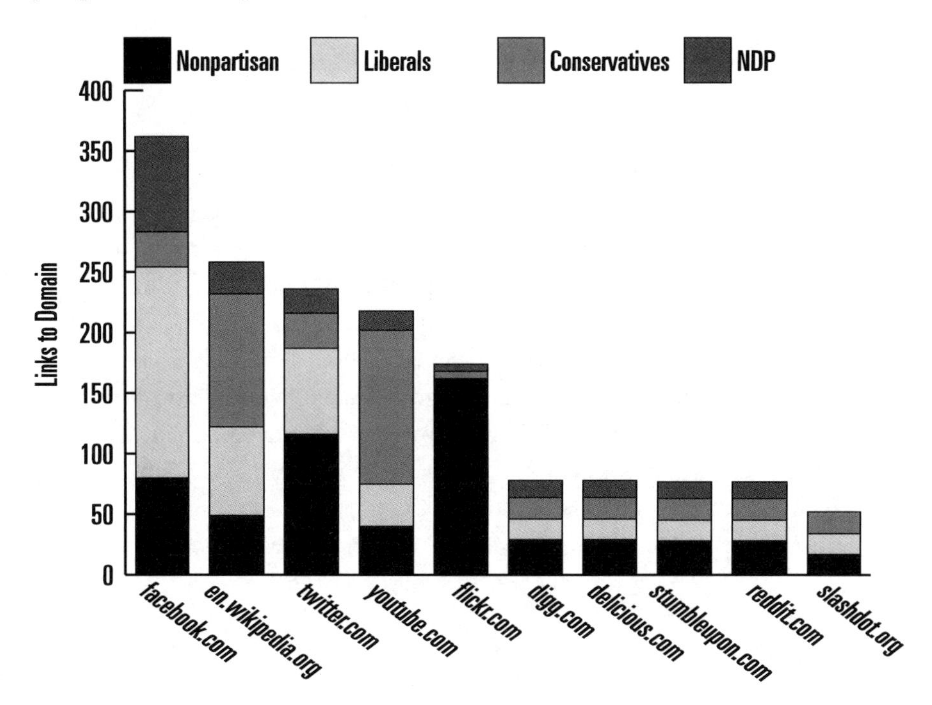

Figure 10. Top "User-Generated Content" and "Online Resources" Domains by Party.

Looking at the specific URLs that were being linked to, it appears that the NDP, Liberals, Greens, and nonpartisans linked to Facebook in order to publicize the online campaign against the prorogation. In terms of links received, the most popular group was the "Canadians Against Proroguing Parliament." By contrast, Conservative links to Wikipedia and YouTube during the crisis had little to do with the prorogation, while most of the Facebook links from the NDP, Liberals, Greens, and nonpartisans went to prorogation-related events or groups. In sum, Conservatives bloggers did not mobilize in support of the prorogation by creating Facebook groups, as was the case during the coalition crisis.

The lack of a pro-prorogation Facebook movement could be interpreted as a deliberate strategy. Surely, if the ruling Conservative Party was worried about negative public opinion, they could have energized their partisan base to support their tactics, as they had done during the coalition crisis. Yet not one single link appears in the data collected to a pro-prorogation group. This suggests that no effort was made on behalf of the party to counter the "Canadians Against Proroguing Parliament" Facebook group. In all likelihood, the governing party reasoned that the 2009 prorogation was presented to the public as a fait accompli: it had been approved by Canada's governor general. It is possible the Conservatives anticipated the crisis would pass as media attention turned elsewhere. Indeed, intense coverage of the Vancouver Olympics ensured that any political bitterness would be quickly forgotten, or at least drowned in the celebratory attention to the major sporting event. Conservative bloggers simply carried about their business as if no crisis was occurring, which in itself could be considered as a specific partisan frame. The NDP, Liberal, Green, and nonpartisan bloggers, on the other hand, focused their energies on supporting anti-prorogation movements on Facebook. Given the success of the antiprorogation Facebook group, this suggests that blogging is no longer the epicenter of new and "more public" forms of political activism online. Blogging mostly deals with the discussion and framing of political issues, whereas new nodes such as Facebook and, more recently, Twitter and text messaging, serve an important role in organizing public movements and protests.[13]

Conclusions

What does hyperlinking reveal about the mobilization of bloggers at different times of the permanent campaign and about the connectivity of the political blogosphere to other Web spheres? Political blogging, at least in Canada, is entrenched along partisan lines. Bloggers supporting the major Canadian political parties tend to be attracted to media outlets that broadly fit with their political views. Thus, while political bloggers might have been seen previously as citizen journalists dedicated to providing a different and more balanced or reasoned voice, the Canadian blogosphere is characterized by its entrenchment along traditional party lines and its predominant reliance on established news outlets rather than alternative or independent media sources or other spheres of influence on the Internet. Further content analyses studies of blog posts and accompanying links would be are needed to further qualify the framing of traditional news sources.

Besides an overreliance upon mainstream news reports and sites, this study of political blogging has concluded that partisanship remains a key organizing principle of the political blogosphere. This was especially clear when looking at links to user-generated content platforms, especially social media networks such as Facebook, where grassroots protests against the Conservative minority government were ignored by Conservative bloggers while they were heavily supported by the Liberal, NDP, Green, and nonpartisan bloggers. As we will see in Chapter Four, political communication and activism on Facebook tends to be vastly different than other online political spheres with regards to which political issues come to prominence and gain users' attention.

Our study investigated how bloggers linked to other spheres. At the time of writing, there have not been, to our knowledge, any studies that investigate the bi-directionality of linking between the blogosphere and other spheres, which would give a more complete picture of not only the role but also the perception of political bloggers online (see Chapter Six for a possible method for cross-platform analysis). We do not yet know the extent to which blog posts are embedded into Facebook wall pages, how blog posts are shared on Twitter or used as Wikipedia sources, etc. Such studies require the development of tools that are capable of correlating different types of hyperlinks—from the traditional hyperlink to the shorter hyperlinks now used on Twitter (bit.ly hyperlinks)—and ones that collect not only hyperlinks but also other sharing protocols, such as those that enable the embedding and sharing of Web objects on different types of platforms.

While the findings presented in this chapter are representative of the English-speaking Canadian political blogosphere and should not be generalized to other political blogospheres, the approach we developed is nevertheless useful for characterizing the broad dynamics expressed in a large sample of blog posts, and is thus useful to define more pointed research questions on subsamples. By focusing on hyperlinking patterns, it becomes possible to see to what extent and when bloggers connect with other political spheres, such as professional politics or grassroots movements. Comparative studies across political blogospheres need to be conducted to see differences in processes of mobilization.

By determining the networked practices of bloggers as a body of mobile partisans, we can gain greater insight into how political parties and the media succeed, struggle, or play surprisingly marginal roles in defining, managing, and dictating the conventions of contemporary partisan politics. Furthermore, this example serves as a helpful point of departure for this book's latter chapters, principally as a means of demarcating the tempo and various modes of partisan participation within a permanent campaign, the links forged between key actors (in this case, bloggers) and social media platforms, and the circulation of influential political objects (blog posts and links). In this way we hope to emphasize the question of networked connectivity and the extent to which the communicative practice of hyperlinking indicates how political bloggers are invoking and involving other documents, objects, and platforms of online political communication.

Last but not least, this chapter focused on political bloggers as exemplars of motorized or mobilized political actors: Canadian English-speaking bloggers are loosely affiliated with a specific party, politically savvy, and partisan. Bloggers, however, represent only a specific type of online political actors. As seen in the last case study in this chapter, other types of actors active on social media platforms other than the blogosphere appear around specific political events such as the prorogation crisis in Canada, which was taken up on Facebook but not in the political blogosphere. This finding indicates that permanent campaigning strategies have to deal with a shifting political terrain: networked political events coalesce on specific platforms and are taken up by different groups of actors. In the next chapter we investigate the Facebook platform further to determine how groups on that platform serve to bring together online publics to discuss and debate specific issues—a networked politics that suggests a much more fragmented political sphere.

Notes

1. See http://nonpartisans.ca/.
2. See http://www.thestar.com/news/canada/article/923225--a-tale-of-two-harpers-five-years-on.
3. A series of graphs and other findings were disseminated by the Canadian Broadcasting Corporation (CBC) as part of a collaborative project between the CBC News and the authors. The project can be accessed on the CBC News website; see http://www.cbc.ca/news/canadavotes/campaign2/ormiston/. Also see Chapter Five for further analysis of research performed during this collaboration.
4. Broken down by blogroll: Blogging Conservatives, 178; Green bloggers, 331; Liberal bloggers, 216; New Democratic bloggers, 117; and nonpartisan bloggers, 127.
5. See http://www.cbc.ca/news/politics/story/2010/01/23/prorogue-protests.html.
6. See http://www.cbc.ca/canada/story/2008/11/26/update-subsidy.html.
7. December 1, 2008 witnessed the release of the coalition government's agreement.
8. See Taylor (2008). Available at stephentaylor.ca/2008/12/announcing-rallyforcanadaca.
9. See http://www.cbc.ca/news/canada/toronto/story/2009/12/30/parliament-prorogation-harper.html.

10. See http://www.cbc.ca/news/canada/story/2010/01/04/facebook-group-prorogation.html.
11. See http://www.cbc.ca/news/canada/toronto/story/2009/12/30/parliament-prorogation-harper.html.
12. See http://www.rideauinstitute.ca/2010/01/21/facebook-anti-prorogation-group-should-not-be-ignored-by-politicians-study/.
13. For example, the role of Twitter and text messaging during the recent political upheavals during the Arab Spring and in European struggles against widespread austerity measures.

References

Anon. (2008). "Ormiston Online." Available at http://www.cbc.ca/news/canadavotes/campaign2/ormiston/.

Bruns, A. (2005). *Gatewatching: Collaborative Online News Production.* New York: Peter Lang.

———. (2008). *Blogs, Wikipedia, Second Life, and Beyond: From Production to Produsage.* New York: Peter Lang.

Bruns, A., Burgess, J., Highfield, T., Kirchhoff, L. & Nicolai, T. (2010). "Mapping the Australian Networked Public Sphere." Paper presented at the Annual Conference of the International Communications Association, June 2010, Singapore.

Drezner, D. & Farrell, H. (2004). "The Power and Politics of Blogs." Paper presented at the Annual Conference of the American Political Science Association, September 2004, Chicago.

Foot, K. A. (2006). "Web Sphere Analysis and Cybercultural Studies." In D. Silver & A. Massanari (eds.), *Critical Cyberculture Studies: Current Terrains, Future Directions* (pp. 88–96). New York: New York University Press.

Garrido, M. & Halavais, A. (2003). "Mapping Networks of Support for the Zapatista Movement: Applying Social Networks Analysis to Study Contemporary Social Movements." In M. McCaughey & M. Ayers (eds.), *Cyberactivism: Online Activism in Theory and Practice* (pp. 165–84). New York: Routledge.

Giasson, T., Raynauld, V. & Darisse, C. (2011). "Hypercitizens from a Distinct Society." *International Journal of Interactive Communication Systems and Technologies* 1 (June), 29–45.

Gillmor, D. (2004). *We the Media: Grassroots Journalism by the People, for the People.* Sebastopol, CA: O'Reilly.

Gleicher, N. (2011). "Moneybombs and Democratic Participation: Regulating Fundraising by Online Intermediaries." *Maryland Law Review* (May), 750–825.

Hackett, R. A. & Zhao, Y. (1998). *Sustaining Democracy?: Journalism and the Politics of Objectivity.* Toronto: University of Toronto Press.

Keen, A. (2007). *The Culture of the Amateur: How Today's Internet Is Killing Our Culture.* New York: Doubleday.

Marres, N. & Rogers, R. (2005). "Recipe for Tracing the Fate of Issues and Their Publics on the Web." In B. Latour & P. Weibel (eds.), *Making Things Public: Atmospheres of Democracy* (pp. 922–935). Cambridge, MA: MIT Press.

McChesney, R. W. (2001). *Rich Media, Poor Democracy: Communication Politics in Dubious Times.* New York: New Press.

Park, H. W. & Thelwall, M. (2008). "Developing Network Indicators for Ideological Landscapes from the Political Blogosphere in South Korea." *Journal of Computer-Mediated Communication* 13 (4), 856–879.

Taylor, J. R. & Van Every, E. J. (2000). *The Emergent Organization: Communication as Its Site and Surface.* Mahwah, NJ: Lawrence Erlbaum Associates.

Tuchman, G. (1978). *Making News: A Study on the Construction of Reality.* New York: Free Press.

CHAPTER 3

Networked Publics: The Double Articulation of Code and Politics on Facebook

The rise of online social networks that enable users to create content, maintain and build social ties, and engage in discussions on public issues has generated much hope—both in the mainstream media and academic circles—for reviving grassroots forms of citizen participation in public affairs. In the past few years, Facebook has become a central component of political activism and campaigning at both local and global levels. The social networking site is seen not only as a tool for managing political discussion online, but perhaps more importantly, as the means through which political communication can be transformed into political action. In comparison to political blogospheres, political activity on Facebook must embrace and cultivate a first-person–framed politics of friendship to enter into the realm of socially networked political action and expression. In other words, in the new Web 2.0 era, and particularly on social network sites such as Facebook, politics is as much about learning the individuated and personal nature of Web software platforms as it is about political communications and content—what we refer to as the double articulation of code and politics. Subsequently, we find that permanent campaigning on Facebook folds social networks in on themselves, producing issue-publics—groups of users coming together to express opinions about particular political issues.

Facebook has emerged as a space where citizens can be informed and mobilized on issues of common interest, come together as a public, and influence political decision-making by engaging in a wide range of institutional and noninstitutional practices—from voting to online and offline protests. This type of description revives the notion of issue-publics (Lippmann, 1922) within the broader theoretical context of the public sphere (Habermas,

1962), and seems to fall in line with the more optimistic discourse equating user-generated content and online social media with democratic participation. While it is undeniable that Facebook makes it possible for anybody with access to the Internet to express themselves, this does not mean that it provides a de facto radical democratization of political activism. Generally speaking, Facebook and social networking platforms cannot be considered as neutral and transparent conduits for a diverse range of political wills. Rather, the logic of the platform itself—from information processing to economic priorities—complicates the formation, organization, and management of what we understand as political activism, and especially the constitution of politically involved publics in various contexts.

This chapter examines the constitution of publics on Facebook as resulting from a "double articulation of code and politics"[1] that creates new conditions and possibilities of political action and communication. We define "double articulation" as the ensemble of processes through which political actors and interests mobilize and invest in code (online platforms, software, networks, informational dynamics, etc.) at the same time as code formalizes and shapes politics (discourses, movements and actors, etc.) according to specific informational logics. From a double-articulation perspective, online publics and issues result from linking, assembling, connecting, and thus hybridizing code and diverse political elements and actors.

This chapter examines the constitution of issue-publics by focusing on three case studies of permanent campaigning on Facebook. We define "networked publics" as those publics who come into being through informational processes. Our exploration of networked publics draws from Maurizio Lazzarato's reflections on how information and communication processes serve to organize a political reality by suggesting, framing, and setting the conditions for the emergence of specific social relations and horizons of political subjectivation (2004). In short, we argue in this chapter that the actors in the permanent campaign are not just investing a political will into the online world; they also have to submit to, adapt to, and make use of the communicative limits and specific informational logics of social media platforms. Political actors emerge not only in relation to an issue, but also in relation to the platform that allows them to come into being. The chapter's first study examines the dynamics of publicity and public discourse on a "me-centric" (social network) platform during the 2007 Ontario provincial election in Canada: the Facebook platform that is always and already personalized for each respective user. No two social networks, and attendant networked contents, are alike—we all see Facebook through our own algorithmically managed social networks. The second case study examines the controversy surrounding the Great Canadian Wish List, a Facebook event launched by Canada's national broadcaster, the Canadian Broadcasting Corporation (CBC), in the summer of 2007. This second study focuses specifically on the strategic use by political actors of the informational logics through which political issues are represented on the Facebook platform. The third case study focuses again on the 2007 Ontario provincial election in Canada, this time examining the critical reconstitution of publics through Facebook's back-end architecture and database, the application programming interface (API). This study addresses the methodological challenges in dealing with black-boxed (closed and proprietary) back-end architectures.

Taken together, these case studies examine processes of permanent campaigning in different political contexts—e.g., election time and time-specific events—and through different platform processes, illustrating the multidimensionality of the shaping of political issues and the mobilization of publics online. These series of double articulations of code and politics highlight the need, from a critical standpoint, to transition away from focusing on the content of online public discussion and to look instead at the modalities of existence for a public and its issues. While the heterogeneous constitution of publics and issues of common interest has been examined through the issue-publics approach (Marres, 2005, 2007), this chapter shows that there is a need to examine not only the assembling of publics, but also the networking of publics via software platforms.

Defining the Site of Analysis: Facebook from a Double-Articulation Perspective

From a social science perspective, understanding the role of social networking sites is a theoretical and methodological challenge. Technically, there are tremendous difficulties in tracking information on private online spaces that have developed complex and black-boxed architectures. Theoretically, the challenges lie primarily in understanding the uniqueness of social networking sites as assemblages where software processes, patterns of information circulation, communicative practices, social practices, and political contexts are articulated with and redefined by each other. In particular, there is a need to examine how diverse elements and actors (human and nonhuman, informational, communicational, and political) are mobilized and articulated in order to shape specific forms of publicity and public discourse. The challenge is that certain elements traditionally ignored by the field of political communication—such as Web 2.0 companies, software processes, and informational architectures—now play a central role in providing the material means of existence for online publics and framing the scope of online political practices. These elements do not simply help transpose a public will onto an online space; they also transform public discussion and regulate the coming into being of a public by imposing specific conditions, possibilities, and limitations of online use. Thus, the challenge is not simply to identify new communicational practices and their effects on the content of public discussion, but to understand how encounters between technologies of communication and political processes create new conditions for the formation of issues of common interest and their publics.

Much hype has surrounded the democratic potential of Web 2.0 platforms as social production tools (Benkler, 2006) to harness collective intelligence, allow users to express themselves by bypassing traditional media (Jenkins, 2006), and enable access to a wealth of information about public issues. The rise of blogs, wikis, and other user-generated content and collaborative platforms has been seen as fundamental in changing the relationships between citizens, politics, and the media (Bruns, 2005, 2008). At the same time, concerns about privacy, surveillance, and control over informational and communicational dynamics have come to temper more optimistic declarations about the renewal of public dialogue and exchange on social networking sites (Albrechtslund, 2008; Boyd, 2008; Petersen, 2008; Scholz, 2008). These concerns highlight the manner in which software and the informational architecture of Web 2.0 sites allow new forms of control to emerge. These new dynamics of control should not be limited to questions of surveillance and privacy, but should also question the cultural experience of being a user of social networking sites. For instance, the informational processes that shape the very experience of being on Facebook—e.g., constant personalization via automated updates and recommendations—are restrictive, as they enable new forms of surveillance and control. They are also productive in that they set the conditions for social bonding as well as political and cultural exchange.

The recent popular uprisings in Arab countries tell a fairly simple story of the role played by social networks in allowing individuals to come together to form a public capable of enacting massive political change. Numerous reports have been circulating in the mass media about protesters in Egypt, for instance, demanding free and unfettered access to Facebook and Twitter as a requisite for building democracy. In these forms of public representations, participatory and social media are not seen simply as modes of enabling public discussion; rather, they are seen as central sites through which political discussion can become political action. Contrary to previous descriptions of the democratic potential of Web 1.0 as a space of open discussion, the democratic claim around Facebook is that open discussion can be mobilized to change actual conditions of existence. In these types of discourses around the political potential of Facebook and other social media platforms there is the assumption that communication technologies are a better and more democratic mirror of grassroots publics. Hence, the conclusion is that Facebook offers tools for better representing a wide variety of issues, some of which are ignored or censored by the mass media, depending on the context. From this perspective, the constitution of publics on Facebook is a matter of individual users gathering together as an accumulation of profiles connected through political groups.

While the Arab uprisings mark a significant turn in that social media appears to be a useful way to promote democratic representation and opens avenues for grassroots political organization, one must be careful about drawing conclusions that new communication technologies create de facto democratic movements and impulses. Ulises Meijias' critique of the "Twitter Revolution" (2008) is a useful reminder that revolutions are about human struggles rather than technological know-how, and that there is a definite pro-corporate, Western-biased ideology at stake in shifting agency from people to communication technologies. After all, the Egyptian protestors were requesting freedom of access to communication without government surveillance or censorship, rather than just access to Facebook or Twitter. It is crucial not to fall into an uncritical position that would see social media platforms, as they exist today, as the ultimate solution for fostering and supporting democratic processes by making it possible for everybody to express themselves. For this reason, it is important to investigate how in North American and Western European sociopolitical contexts, such as those that enable and even encourage participation on social networks as a way to revive public life and community, Facebook is not a neutral space for aggregating publics. Rather, it articulates different political-economic interests along with specific informational logics. One should not ignore, for instance, the growing use of social networking sites by political parties and political figures that was epitomized during the 2008 Obama campaign in the U.S. presidential election. In this case, the use of decentralized networks of users was key in both garnering widespread support and raising funds. Thus, the terrain of politics is not so much divided between traditional politics (i.e., electoral campaigns) and radically new and democratic politics on social networks; rather, we can expect to see efforts to link and bridge differing political logics and interests by making use of the communicative capacities and informational logics of social networking platforms. Thus, the ongoing debate between optimistic and pessimistic descriptions of participatory media highlights the need to redefine objects of study—such as online political discourses—not simply as the work of human actors (e.g., citizens, journalists, politicians) in specific political contexts, but also as moments of encounter with an informational structure that, to a great extent, organizes the parameters of political communication and action online. Paying attention to the informational architecture of participatory media makes it possible to understand the dynamics and nascent techno-political strategies through which some political discourses and action online become more prominent than others.

From a methodological perspective, online participatory media sites should be seen as platforms: the convergence of different technical systems, protocols, and networks which enable specific user practices and connect users in particular ways (McKelvey, 2008). There is a need to follow a vertical approach (Elmer, 2006) to study the conditions of connectivity of different cultural, communicational, and informational components present in specific online spaces and platforms. The code, languages, and architectures, as well as the other elements that produce a human-understandable visual interface, impose specific constraints on communicational process while also allowing for new possibilities of expression. As such, they redefine what it means to communicate online. With Facebook, a vertical approach that takes into account the "informational politics" (Rogers, 2004) regulating communication processes leads to an examination of the paradox of "free and open communication" that exists between the use of tools to facilitate the production and circulation of content and the opacity and complexity of an architecture regulated by the economies of data-mining. The vertical approach invites us to examine the double articulation of code and politics and to identify how informational processes intervene in the plane of politics and public discussion, and vice versa.

The Aggregation of Publics on Me-Centric Platforms: The 2007 Ontario Election

What are the informational specificities of social networking sites such as Facebook? Social networking sites (SNSes) share some similarities with other Web 2.0 platforms, but differ significantly in terms of patterns of information circulation. SNSes allow people to be in touch with their social circle, and typically begin with a request to create a personal account. From their personal account pages, users can invite other users to become "friends" and can send messages and content (text, video, pictures, music) to their network of friends. On Facebook, the modalities of exchange among friends are extremely varied, from news stories that automatically let users know about the Facebook activities of their friends to private messages and public (e.g., "wall-to-wall" posts) exchanges. One's network can be extended not only through invitations to friends, but also by becoming, for instance, a fan of a public figure, political cause, or TV series, and also by creating and/or signing up for events and groups.

As opposed to other Web 2.0 spaces, social networks can be considered first-person, or "me-centric," social media platforms that are entirely produced, formatted, and visualized through individual users' own personal and unique social networks. While Facebook aims to act as a large repository of information by offering numerous ways to post and access information and to communicate with others, any kind of activity on this platform takes place through a highly individualized and personalized perspective. The entry point on the Facebook interface is one's user account, and Facebook's recommendation and search features rank their results by measuring closeness to one's network, including school affiliation, geographic location, and the number of friends already participating in an event or group. The me-centric perspective on Facebook is therefore a constraint. However, it also offers a specific context for cultural and social experience and enables new forms of sociality and access to information. In the same way, the informational processes aggregating users into groups define specific modalities of publicization.

Modes of political activity on Facebook differ significantly from other Web 2.0 formats, especially with regard to how information and content circulate through the Facebook platform. By default, Facebook provides three ways to express political support: (1) members can become "fans" or supporters of a politician's profile; (2) they can create or join a group; and (3) they can state their political views on their profile. Of these three modes of expression, only groups and politicians' profiles extend a person's political views beyond his or her me-centric network to publicly list the person as a member or supporter on the Facebook interface. Becoming a fan or a group member allows for the aggregation and publicizing of users around particular political figures and causes. In this way, networks of common interests are created, and the Facebook platform greatly simplifies communicating with a network of users via emails, invitations, and reminders. In turn, users can invite their own friends to join a group, page, or event, either in an active fashion (e.g., sending a message) or in more passive ways (e.g., through automated status updates visible to friends). Facebook groups offer a far more dynamic window of study—one that reflects the participatory aspects of social media—than a politician's profile page, which comes only from top-down decision-making processes by parties and campaigns. Only politicians can create political profiles, whereas anyone can create or join a group, thus offering broader avenues for political participation. For these reasons, exploring the formation of participatory publics on Facebook is best done through a study of groups, as any user can create a group, and groups represent the moment at which me-centric profiles are assembled and publicized. Furthermore, the logic of Facebook groups is that of making one's preferences as a user consciously public, as opposed to maintaining some level of privacy with one's personal profiles and other more intimate forums of communication within one's networks of friends (including updates and wall-to-wall communication). Facebook groups have been made public by Facebook so that they are now listed on search engines, thus making data collection simpler.

How does the coexistence of the two poles of me-centricity and publicity on Facebook shape the constitution of publics around issues of common interest? First, it must be acknowledged that the informational structure of representation on Facebook is not a neutral translator of political will. Furthermore, Facebook should not be considered a stable format. It is a platform that allows for different modulations of me-centricity and publicity. Therefore, the first set of double articulations to consider concerns: (1) the main actors involved in the constitution of issues and their publics—the politicians, citizens, NGOs, etc. that make use of the communicative possibilities offered by Facebook in order to support, represent, and create new modes of political communication and participation; and (2) the new communicative affordances offered to users by a software platform, and the ways in which they allow (and also disallow) new modalities of political discussion and political action.

Campaigning on Facebook Groups

The 2007 Ontario provincial election provides a useful case study for looking at how this first set of double articulations highlights the new political dynamics of publicity on me-centric platforms. When undertaking this research project, our first intention was to see if it would be feasible to track the rise of issues of common interest and the constitution of their publics by taking into account the specific communicational dynamics of Facebook. We aimed to see if Facebook was mobilized by traditional political practices such as partisanship, and if it allowed for new formations, issues, and publics to emerge. Our method involved weekly manual searches for groups on Facebook, using the names of the major party leaders "Howard Hampton" (New Democratic Party), "John Tory" (Conservative), and "Dalton McGuinty" (Liberal), as well as the more general search term "Ontario election 2007." We included the last category as a catchall for issues related to the electoral campaign. In total, 281 Facebook groups were tracked from August 31, 2007 to October 5, 2007. The groups ranged from official party groups to local groups and grassroots groups.

Overall, we found a coexistence of partisan and issue-based politics for the Ontario election. While the majority of Facebook groups (54 percent) were used for partisan purposes in that they either supported or attacked a candidate representing a particular party, there was also a significant number of groups focused on issues of common interest (38 percent of all groups). In that sense, there was a rearticulation of a campaign on Facebook as a way to pressure politicians into acting on specific issues, such as lowering or freezing university tuition fees or raising the minimum wage. It is also notable that the most popular groups in terms of number of members were almost all issue-based rather than party-based (see Figure 1).

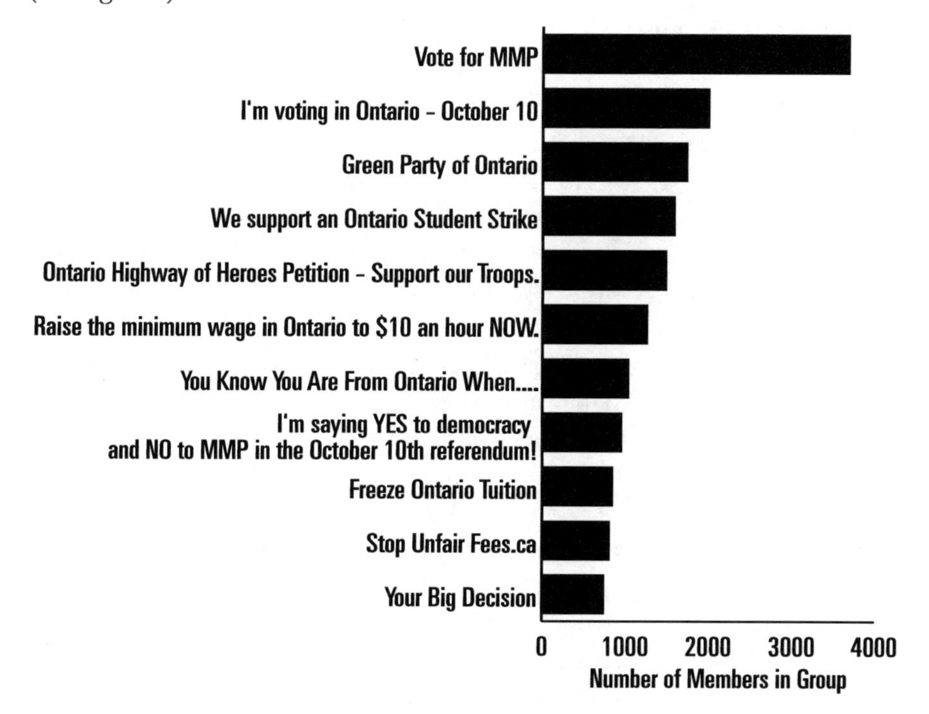

Figure 1. Top Ten Facebook Groups by Membership for Ontario Election, October 5, 2007.

The ranking of the top ten Facebook groups by membership for the last week of the campaign, for instance, showed that the "Green Party of Ontario" was the only group in support of a political party. The rest of the groups were focused on issues, from electoral reform ("Vote for MMP") to tuition fees ("Freeze Ontario Tuition") and renaming a highway ("Ontario Highway of Heroes Petition—Support Our Troops"). A common conception of participatory media is that they enable marginalized voices to express themselves. Indeed, in the case of the Ontario election, Facebook allowed for the emergence of marginalized issues and publics. Some of the most popular Facebook groups during the election (in terms of number of members) focused on issues that were not prominent in the traditional mass media. The Green Party of Ontario, for instance, had a strong grassroots and official presence on Facebook ("Green Party of Ontario," 1,736 members). Throughout the campaign, another prominent Facebook group was one related to a referendum on electoral reform taking place at the same time as the provincial election. The issue of electoral reform was widely debated on Facebook but mostly ignored by the mass media, which focused more on the politicians themselves (Nelson, 2007).

When looking at the aggregation of publics on Facebook one must avoid the assumption that groups with the highest number of members are the most effective ones. While it is undeniable that groups with high membership act as central nodes for political discussion and decision-making, the communicative affordances of Facebook—along with its me-centric architecture—reshape a horizon of political agency within more localized and personalized networks. During the 2007 election there were numerous local groups focusing on similar issues, thus showing that Facebook can be used not only to publicize issues, but also to foster a process of satellization of issues within smaller personal networks. For instance, twenty-four groups related to the proposed electoral reform were created during the Ontario campaign. Eight of these groups were against electoral reform, two were neutral, and fourteen were in favor. While the group "Vote for MMP" reached the largest membership, with 3,685 members on October 5, 2007, there were only five other groups about electoral reforms with more than 100 members. The other eighteen groups ranged from one to seventy-eight members. An effect of satellization is thus created in that small groups echo the same message (i.e., "Vote 'Yes' for Ontario Electoral Reform" and "Vote YES!!! to Election Reform"). The rearticulation of the process of publicity through the personalizing and me-centric informational dynamics of Facebook reshapes our conception of the public from a large body of concerned citizens to smaller aggregations of users.

As we've already established, political activity on Facebook cannot be treated as a popularity contest where the groups with the most members are seen as the most important or powerful. Rather, it is important to understand the different communicational logics at stake in groups with large memberships who have their common interests expressed by joining a particular group, or smaller aggregates of users whose membership in a particular group is but one of the many commonalities they share. More research into patterns of group membership needs to be done, and further analysis has to be developed into the political and strategic importance of these different types of groups. A hypothesis that needs further exploration is that groups with large memberships can be used to make an issue more visible to the mainstream media and professional politicians, whereas small satellite groups might play an important role in convincing and mobilizing voters, perhaps akin to the information-leader model developed by Laswell's two-step flow of communication. Overall, in our studies of the 2007 Canadian electoral campaigns, it appeared that the Facebook platform enabled new kinds of participation through the double articulation of tools for participatory communication and underrepresented publics and issues and, at the same time, personalized informational networks and issues of common interest.

Reifying a Public: The Great Canadian Wish List

While an analysis of Facebook groups can take place at the level of the articulation of me-centric architecture and publicity, through another Canadian case study ("the Great Canadian Wish List"), we would like to highlight the importance of studying the double articulation of code and politics through the transposition of issues and publics into the specific communicational and informational regimes of the Facebook platform. It is not simply a question of the aggregation of publics on Facebook, but also cultivating an awareness of the hidden power dynamics that play an essential role in defining which publics can come into being.

In the summer of 2007, the Canadian Broadcasting Corporation (CBC) launched a Facebook campaign to uncover the top wishes of Canadians. In its attempt to link together citizens, traditional mass media, and online social networks in a novel way, the CBC invited Canadians—whether living in Canada or abroad—to join the Great Canadian Wish List group on Facebook to see what "Canadians hope for the country's future" and to define their own wishes (CBC, 2007). The platform enabled those who joined to express their support for one or more wishes by becoming friends with a wish: "Facebook will automatically rank the wishes: the more 'friends' your wish has, the higher its place" (CBC, 2007).

The premise of the Great Canadian Wish List was to use Facebook as a tool for democratic and participatory communication that would enable a broad public to define their own issues. However, the campaign illustrates how processes of double articulation can take place at the intersection of informational dynamics and the representational economies of the user interface. The user interface is not only about content and discourse but also about representing and translating political actions that are mediated by informational practices. A click of the mouse, for instance, is equated with a political act, and represented as a vote or an agreement in the case of an online petition, but this equation quickly becomes problematic because the conditions of political participation are affected by the different technologies of communication used. The Great Canadian Wish List underlined the double articulation that connects informational dynamics with the representational interface, and the critical importance and limitations of the user interface as a mode of existentializing and reifying informational processes as cultural practices.

With regards to fulfilling its mandate of representing Canada to Canadians,[2] the Great Canadian Wish List was an attempt to bypass traditional filters in order to enhance direct communication among citizens. The event was linked directly to a mass media system, as the final wish list was announced on TV, radio, and the Internet. The Great Canadian Wish List was an experiment in articulating a political practice—voting for a cause—with a specific informational practice: "friending" a group on Facebook. This articulation, however, led to a redefinition of what it means to be an active citizen, especially as the Facebook architecture does not allow broad public discussion on groups. In order to post a comment, video, or photo to a group, one has to be a friend of that group. Thus, the Great Canadian Wish List delineated the discursive agencies of members of different publics by imposing a structural barrier on the horizon of citizenship.

The Great Canadian Wish List was a notable mass media event because of the controversy it raised, which revealed a second process of double articulation where pre-existing publics intervene in the equation between the communicational practice of "friending" on Facebook and the political vote. At the end of the campaign, the top wish in terms of number of friends was "Abolish Abortion in Canada," launched by pro-life activists. The second most popular wish was that abortion remain legal. This example—of how a pre-existing network of interests (the anti-abortion movement) could push for an issue online and have it reified as a "Canadian wish" representative of all Canadians—reveals how political strategies can make use of informational dynamics (such as "friending") in order to legitimize an issue.[3] This type of strategy, known on the Web as "freeping," originates from the calls by members of the conservative, U.S.-centric Free Republic forum to influence online polling by voting en masse.[4]

The Great Canadian Wish List revealed a new set of informational politics intervening in the process of constituting and legitimizing issues and their publics. The final wish list showed that far from representing the wishes of everybody, or even representing wishes corresponding to the demographic profile of Facebook users, the mediation of a large-scale communication campaign undermined the cultural assumption that ease of participation via online tools leads to balanced representation and democratic communication (Canadian Press, 2007; *Montreal Gazette*, 2007). The Great Canadian Wish List experiment highlights that what at first seems like a simple transposition of politics onto a social networking site actually requires the alignment and stabilization of a series of informational, political, and representational processes. In the case of the Great Canadian Wish List, the possibilities of intervention at the level of the articulation of informational dynamics—number of votes, number of clicks—with representational processes—participatory media as democratic representation—highlights the growing importance of these new techno-political practices. Thus, examining the constitution of issues and publics on SNSes requires taking into account the unique techno-political possibilities and horizons offered by different online spaces.

Opening the Black Box: Critical Reconstructions of Publicity

The question of political involvement on Facebook necessitates reflection on the different mechanisms aggregating publics around specific issues, and on the reification of issues and their publics through the information-ranking logic of platforms. From this, it becomes apparent that the concept of political representation has to be understood as strategically mediated by informational logics. The user interface provides a rich set of processes to be analyzed, whether by looking at content as expressive of participatory communication or by looking at the processes through which politico-informational dynamics are represented and therefore reified and legitimized on a platform. While the interface is an important site of analysis, it offers a limited perspective in that the informational dynamics that shape the interface are invisible to individual users and, to a large extent, also communications researchers. Facebook is a black box (Latour, 1987, p. 131), and therefore it is difficult to examine some of the opaque techniques at the informational level that could participate in shaping specific representations of users and political processes. In the final case study, also conducted in the context of the 2007 Ontario provincial election campaign, we highlight how it is possible to critically highlight and reconstruct informational and political practices that are not visible at the interface level, yet are central to understanding processes of political affiliation and support.

Facebook epitomizes the paradox of the representational interface: It reveals a specific range of communicative possibilities while hiding and rendering invisible some of the core informational processes that might reveal something about the ways in which we perceive ourselves and others as members of a public, and how we experience our social and political world online. Wendy Chun (2005) describes software as ideological. The ideological component to Facebook is that its interface hides the technical elements and informational processes that shape political practices and the horizon of political subjectivation. It is interesting to note that, for instance, the Web 1.0 standard of the site map as a means for individual users to get an overview of the informational space of a website does not exist on Facebook; because the platform is geared toward me-centricity (personalization and customization), there can be no third-person, objective worldview. Being forced to put on the blinders of me-centricity is a challenge for communications research, especially with regards to studying issue-publics from a limited first-person horizon. Opening the Facebook black box, or at least understanding its functioning, can expose the social and cultural assumptions embedded within specific informational dynamics and also allow for a rearticulation of our online political horizons. This would allow researchers and citizens to construct modes of critical inquiry that challenge the me-centric perspective.

As an illustration of the limits of interface-based research and the need to develop an approach focusing more on the hidden informational dynamics of Facebook, during the 2007 campaign for the Ontario provincial election we conducted a series of experiments related to group membership that involved using Facebook's application programming interface (API). APIs enable connections between different software and, in the case of Facebook, between third-party software applications and the Facebook databases that contain information about users, events, groups, pages, etc. The Facebook API is a central component of the platform in that it allows for the production and distribution of the many applications developed by third parties—it is here that third-person metrics lie. A political application, such as the one that enabled users to share the famous "Yes We Can" song in support of Barack Obama's campaign for the U.S. presidency in 2008, enacts a series of double articulations of code and politics that differ from the ones observed in Facebook groups. On the one hand, the specific informational dynamics for sharing information on Facebook are mobilized for a political cause: in this example, spreading a political anthem to large audiences. On the other hand, the code is not simply mobilized by political forces; it also enables new forms of covert political practices that make use of the surveillance potential of Facebook to gather information about subscribers to this application—their demographic profiles, likes and dislikes, networks of friends, etc. The assemblage that results from the deployment of this specific Facebook application should not be considered as a homogeneous whole, but rather as allowing for multiple power relations—some of which are visible, while others remain hidden. Because online social networks such as Facebook are layered entities (Langlois, 2005) that involve not only visual interfaces but also informational processes and communicational practices, these double articulations take place on different planes with varying effects. There is a need to develop analytical approaches and methodologies that move away from the user interface as a central source of data and to integrate the informational dynamics that, while invisible to users, play a central role in defining the modalities of existence for publics on Facebook.

In the process of collecting groups during the 2007 Ontario provincial election, we noted that one of the blind spots of Facebook's user interface was that it did not enable the visualization of links between groups. This is something we wanted to investigate because it would answer the question as to whether members of different publics share common issues. Typically, lists of groups are suggested to a user on a group page, but these are personalized to one's network. One way of visualizing how groups are linked together (or not) is to see which members subscribe to more than one group. While this information is not visible via the first-person–designed Facebook user interface, it is accessible via the Facebook API. We collected the user IDs of the Facebook groups related to the 2007 Ontario election via the Facebook API, and identified which users were subscribed to more than one group. We then asked the API for the users' names and identified which users were public political figures. Using the network visualization software Réseau-Lu,[5] it was then possible to make this network of connections visible (Figure 2).

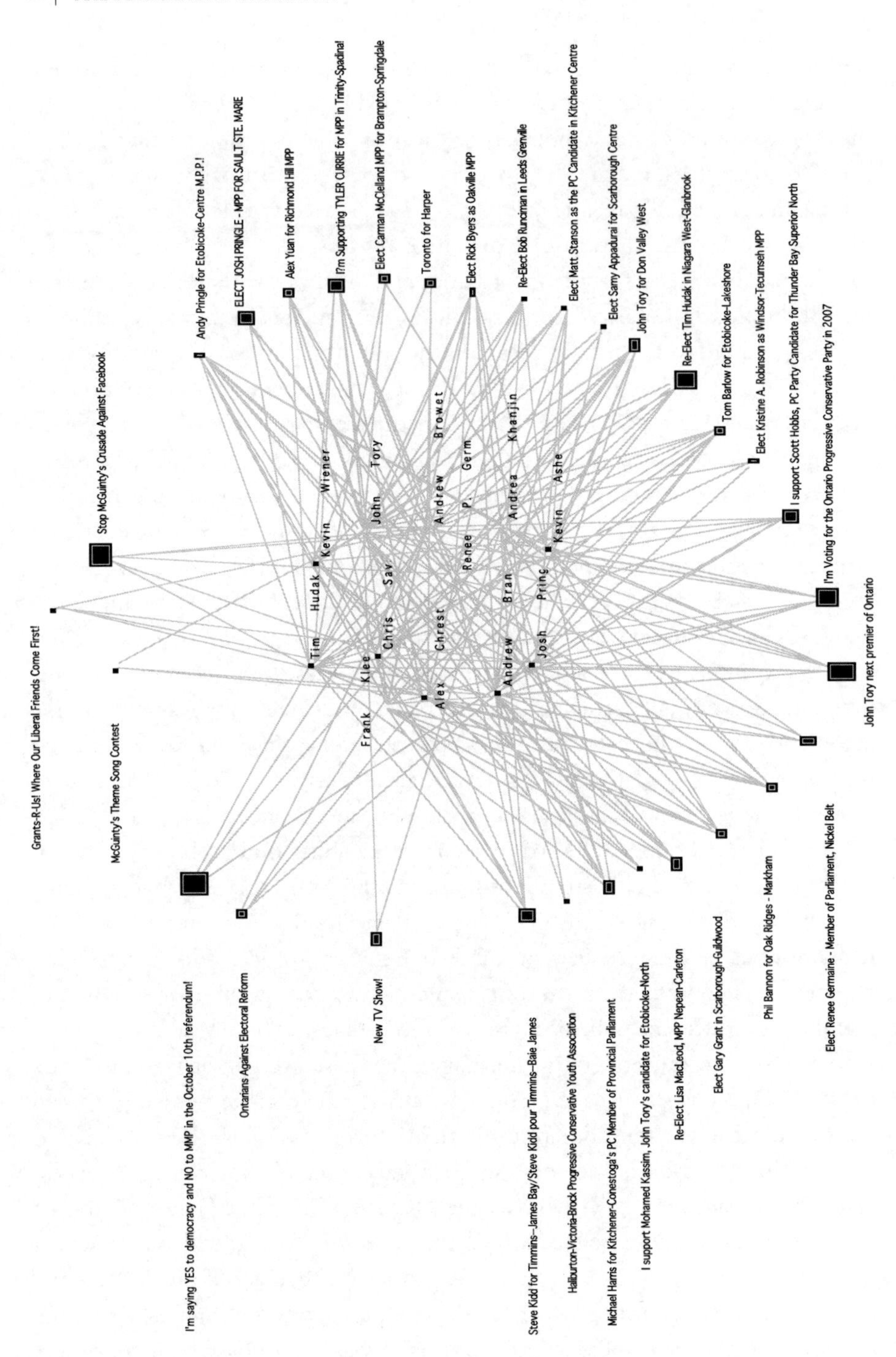

Figure 2: Ontario 2007 Election—Group Members Affiliated with More than One Facebook Group.

Through these visualizations, it was possible to see which politicians have strategized their Facebook presence—that is, which politicians subscribed to more than one group—and whether there were clear patterns for subscriptions according to party affiliations. In the case of the Ontario election, it became clear that members of the Conservative Party strategically subscribed to pro-Conservative groups as well as to anti-Liberal groups. The strategy of supporting one's party while undermining the party in power became apparent through the visualization software, and revealed the extent to which groups have to be studied not only in terms of their content but also in terms of their location within strategic networks.

This experiment illustrates the possibility of critically assessing the constitution of publics according to informational networks while bypassing the user interface altogether, and of constructing other interfaces to reveal invisible informational and communicational processes that nevertheless speak to how users build their membership among diverse publics and issues. In this particular instance, we were looking at political performativity rather than discursivity; that is, how users themselves adopt specific informational practices that reflect political strategies. From a research perspective, tapping into the potential of the Facebook API is fraught with ethical and terms-of-use issues. For instance, it is relatively easy to obtain the Facebook identification of group members, and from there, to ask the Facebook API information about these users, although this could be considered a breach of privacy and against Facebook's terms of use. The problem with Facebook is that it has developed a model where there is no difference between online and offline identity. This is something that became crucial for the Egyptian government's ability to identify and crack down on political dissenters. Without question, there needs to be more sustained reflection on the problematic ethics of Facebook's lack of anonymity. One possibility could be for researchers to develop their own Facebook-approved application that invites users to subscribe to an application that would track and anonymize their political activity on the platform. In this way, researchers would be able to act according to research ethics principles and in compliance with the Facebook API's terms of use.

Issue-Publics and Networked Publics

If social networking sites—and Facebook in particular—have emerged as the platforms of expression and action for publics, how can we actually examine the constitution of these publics and their involvement in different types of issues? The three case studies discussed in this chapter show that the constitution of online publics involves examining how political processes, issues of public interest, and public discussion and action take part in a double articulation of code and politics; that is, how they are reshaped within specific communicational spheres that employ specific informational architectures. On Facebook, these double articulations take place within a me-centric and public continuum that involves three different layers: (1) the user interface; (2) informational dynamics; and (3) hidden back-end networks. The three case studies present some important methodological and theoretical issues relating to the shift from looking at online politics primarily in terms of the content of political discussion, to focusing on the heterogeneous (informational, performative, discursive) processes of assembling and connecting publics with specific issues.

Noortje Marres' work on issue-publics offers a theoretical starting point for further reflection on the constitutive aspects of assembling issues and their publics. According to Marres (2007), interest in the constitution of heterogeneous networks by which a public issue comes into existence, and the processes by which it is addressed by members of a public, merges science and technology studies (STS) with the pragmatist tradition of John Dewey and Walter Lippmann. Marres argues that a public is constituted as a heterogeneous assemblage that comes into being through the process of defining the scope of an issue, or problem, of common interest (2005, 2007). Invoking Dewey and Lippmann as well as actor-network theory (ANT), Marres contends that the definition of an issue and the modalities of public participation arise from specific "conjunctural dynamics" (2005, p. 212). Also of particular relevance here is Bruno Latour's (2004, 2005a, 2005b) examination of how issues bridging scientific, political, and social dynamics (e.g., global warming and genetically modified foods) have been ignored by social and political sciences. Latour argues that analysis of the procedures of democratic deliberation is too often myopically focused on either discursive or formal elements, and that there is a tendency to overlook a public's involvement in an issue beyond the existing institutional apparatus for political decision-making. Attention must be paid to how the definition of an issue includes not only human actors but also nonhuman elements, such as the objects and techniques that organize public responsibility; how different groups (of politicians, citizens, NGOs) fight over the boundaries of an issue; and how these participations and affordances are shaped through specific informational practices and communicational processes.

The issue-publics concept presents a strong framework for looking at the double articulation of code and politics in the case of assembling publics on

Facebook, and potentially, the Web at large. Including technologies as active participants in the constitution of issues and their publics creates new avenues for examining how Web-based communication technologies intervene in the political process. The issue-networks concept (Rogers, 2004; Marres & Rogers, 2005) was developed as a way of examining the unfolding of issue-publics on the Web and analyzing online politics as a form of political action. The tracing of issue-networks involves examining how different actors (e.g., NGOs, political parties, etc.) involved in an issue of common interest can debate and define the scope of the issue and the relationships among themselves through hyperlinking patterns. The analysis of issue-networks traditionally involves following the evolution of a hyperlinked network—of a range of websites related to a common issue—to understand the dynamics among different political actors. In this sense, an issue-network can be understood as a "heterogeneous set of entities (organizations, individuals, documents, slogans, imagery) that have configured into a hyperlink network around a common problematic, summed up in a keyword such as climate change" (Marres & Rogers, 2005, p. 928). The relationship between issue-publics and issue-networks on the Web is based on the assumption that the Web functions as an archive that enacts traceability (Marres, 2005, p. 109). Thus, online communication practices (e.g., hyperlinking) are seen by Marres as enacting the unfolding of issue-publics and the Web becomes a "site for the performance of a controversy" (p. 109). The previous chapter on political blogging was inspired mostly by such a hyperlink method, as the contours of bloggers as political and media actors were defined by analyzing their hyperlinking patterns.

The transition from issue-publics to issue-networks is much more problematic in the case of online social networks. The first set of problems is methodological, in that the shift from Web 1.0 to Web 2.0 has undermined both the centrality of hyperlinks as the organizing principle of online communication and the status of the Web as a relatively open and traceable archive. SNSes such as Facebook are characterized by their enclosed, portalized structure that invites users to stay within the website and forces them to explore the site through limited me-centric perspectives. As researchers, we found this informational structure particularly frustrating, especially upon discovering that search returns will always differ from one user to the next because of the constant personalization at stake with Facebook. Furthermore, the linking of information on SNSes relies on new languages, protocols, and practices. With this new configuration, traditional hyperlinking becomes only one informational tool among many, thus making it impossible to use the same methods of data collection that were developed for the Web 1.0

environment. We concluded from our case studies that the issue-networks approach needs to be reconceptualized and expanded so that it takes into account the specificities of Web 2.0 platforms. This does not mean simply incorporating other protocols that enable the circulation of content on Web 2.0 platforms (as exemplified in Chapter Six), but also finding ways to address the growing black-boxing of Web 2.0 spaces and the ensuing difficulties with tracking and recording content. Platform-specific methodologies that go beyond the interface level and tap into the data collected by the platform are promising yet challenging sites of analysis for communications research. While the Facebook API can give more clues as to what are the unrepresented practices by members of a public, the use of the Facebook API requires sustained consideration of issues surrounding the privacy of Facebook users. All these challenges point toward the need for further research into platform-specific methods. As researchers struggling to break open the black box, we hope that other researchers will help to develop these tools.

As our case studies illustrate, the issue-networks approach needs to be thought of not only in horizontal terms (the expansion of a hyperlink network) but also in vertical terms. The different layers and processes involved at the back end might allow for the exploration of dynamics that are not visible at the level of the user interface but nevertheless play a central role in regulating and allowing for new forms of intervention. Greater attention must be paid to the dynamics of the online informational milieu (Terranova, 2004, p. 52) in order to understand the micro and macro changes that take place in the transition from the political to the techno-political. In this new configuration, the methodological challenge lies in identifying the many sites and layers of articulations of code and politics, and in tracing the movements—the new trajectories of power and knowledge—that link informational dynamics with political and research practices in specific contexts.

At the theoretical level, the assumption that online communicative practices enact, transcribe, or represent the constitution of issue-publics in the case of online social networks is not tenable. The double articulation of code and politics—in the case of SNSes—is not an unproblematic translation of one dynamic onto another communicative plane. The rise of software as a personalized, first-person actor and Web 2.0 platforms as new types of techno-cultural spaces fundamentally changes the constitutive dynamics of issues and their publics. As Mejias (2008) points out: "Networks—as assemblages of people, technology and social norms—arrange subjects into structures and define the parameters for their interaction, thus actively shaping their social realities. But what does the social network include, and what is left out?" Informational dynamics cannot be seen as supports of communicational dynamics anymore, but as direct interventions in the communicational, cultural, and by extension, the political plane.

The issue-publics approach would benefit from Lazzarato's exploration of the role played by informational labor in control societies. Drawing from Michel Foucault, Gilles Deleuze, and Gabriel Tarde, Lazzarato looks at the constitutive relations between human and nonhuman actors. In particular, he examines the diverse elements and fluxes (technological, psychological, economic, political, etc.) that modulate, regulate, and condition such relations according to specific power interests. The advantage of Lazzarato's approach is that it redefines control as the management of the ensemble of processes and conditions that make specific situations real and visible. This approach also invites us to acknowledge that the primary role played by online social networking sites—particularly private and commercial ones such as Facebook—is to give us our horizons; our sense of the possible. No longer is the content of a social network a sufficient descriptor; instead, we must treat SNSes as locations of existence with their own limits of possibility. In the same way, Facebook should be seen as creating the horizons, tools, and practices which—through the double articulation of code and politics—give birth to new social actors, new modes of agency and subjectivation, and new limitations and power relations. The tracing of issue-publics in their multidimensionality on SNSes can give way to a critical assessment of SNS politics.

Lazzarato's work on immaterial labor is useful not only for examining the assembling of issue-publics writ large, but also for this specific moment of transition when an online public is networked. The online informational systems provide the material, communicational, and social means for a public to exist, and this takes place through the implementation of a network that defines the parameters for the agency of a public and its specific communicative affordances. In this sense, the network provides the parameters for assembling issues and their publics in specific ways. As seen in the second case study, the network can impose a specific communicative discipline, and at the same time, it can offer possibilities for the rearticulation of pre-existing power dynamics. Analysis of the networking of a public, as demonstrated in the third case study on group membership patterns, makes it possible to envision interventions—from a critical research perspective—into the processes of online public participation by revealing unrepresented dynamics. In terms of rebuilding a critical approach to the democratic potential of Web 2.0, analyzing the networking of publics and issues finds some resonance with Chris Kelty's exploration of the open-source movement as a "recursive public," which he defines as a "public that is vitally concerned with the material and practical maintenance and modification of the technical, legal, practical and conceptual means of its own existence as a public" (2008, p. 3). A more critical approach to politics on social networking platforms should include awareness of the networked means of existence of a public, making it possible to trace the new power dynamics in the shaping of issue-publics that arise through the double articulation of code and politics. Such an analysis—as demonstrated through the case studies discussed in this chapter—is rife with practical, methodological, and theoretical challenges. Collaborative efforts across disciplines, therefore, need to take place in order to develop the tools and theoretical understanding of the different, concurrent, and sometimes paradoxical modalities of Facebook, SNSes, and Web 2.0 sites in general.

The rise of social networking sites as platforms for political communication and action thus presents researchers with a complex set of issues regarding the double articulation of informational logics and political will, of communicative capacities and political expression, and of algorithmic logic and political representation. Social networking sites allow for the rise of a new type of actor: the aggregated publics that come into being around specific sets of issues and attempt to change the political agenda by making their issues more visible, using the informational logic of the platform. In order to fully examine the role of such Facebook-enabled issue-publics in the permanent campaign, we have concluded that the communicative logics, capacities, and limits of the platform must be taken into consideration. In particular, in the context of permanent campaigning and its role in software-enabled forms of politics, we've concluded that the effects of first-person interfaces and political frames—or conversely, access to Facebook's third-person API data—largely defines the architecture of contemporary networked politics, the populist possibilities, and the institutional means of controlling the political agenda.

Notes

1. Deleuze and Guattari (1987).
2. See http://cbc-radiocanada.ca/about/mandate.shtml.
3. Dave Gilbert, the creator of the anti-abortion wish on Facebook, has admitted himself that the results from the Great Canadian Wish List do not constitute an official pc show that the potential for further mobilization exists. See http://www.thei aug/14abolishabortion.html.

4. See http://en.wikipedia.org/wiki/Free_Republic.
5. Réseau-Lu was designed by the consulting and software development company Aguidel Consulting, Paris.

References

Albrechtslund, A. (2008). "Online Social Networking as Participatory Surveillance." *First Monday* 13 (3). Available at http://www.uic.edu/htbin/cgiwrap/bin/ojs/index.php/fm/article/view/2142/1949.

"Announcement: Facebook/ABC News Election '08." (2008). Available at http://www.pressreleasepoint.com/announcement-facebookabc-news-election-03908.

Benkler, Y. (2007). *The Wealth of Networks: How Social Production Transforms Markets and Freedom.* New Haven, CT: Yale University Press.

Boyd, D. (2008). "Facebook's Privacy Trainwreck: Exposure, Invasion and Social Convergence." *Convergence* 13 (1), 13–20.

Bruns, A. (2005). *Gatewatching: Collaborative Online News Production.* New York: Peter Lang.

———. (2008). *Blogs, Wikipedia, Second Life, and Beyond: From Production to Produsage.* New York: Peter Lang.

CBC. (2007). "Canadians Hope for the Country's Future." Available at http://groups.yahoo.com/group/IFNEWS/message/2212.

Chun, W. (2005). "On Software, or the Persistence of Visual Knowledge." *grey room* 18 (5), 26–51.

Deleuze, G. & Guattari, F. (1987). *A Thousand Plateaus: Capitalism and Schizophrenia.* Minneapolis: University of Minnesota Press.

Elmer, G. (2006). "The Vertical (Layered) Net." In D. Silver (ed.), *Critical Cyberculture Studies: New Directions* (pp. 159–167). New York: New York University Press.

Habermas, J. (1962). *The Structural Transformation of the Public Sphere: An Inquiry into a Category of Bourgeois Society.* London: Polity Press.

Jenkins, H. (2006). *Convergence Culture: Where Old and New Media Collide.* New York: New York University Press.

Kelty, C. (2008). *Two Bits: The Cultural Significance of Free Software.* London & Durham, NC: Duke University Press. Available online at http://twobits.net/pub/Kelty-TwoBits.pdf.

Langlois, G. (2005). "Networks and Layers: Technocultural Encodings of the World Wide Web." *Canadian Journal of Communication* 30 (4). Available at http://www.cjc-online.ca/index.php/journal/article/view/1636.

Latour, B. (1987). *Science in Action: How to Follow Scientists and Engineers through Society.* Cambridge, MA: Harvard University Press.

———. (2007). *Reassembling the Social: An Introduction to Actor-Network-Theory.* Oxford, UK: Oxford University Press.

Latour, B. & Weibel, P. (2005). *Making Things Public: Atmospheres of Democracy*. Cambridge, MA: MIT Press.

Lazzarato, M. (2004). *Les Révolutions du Capitalisme*. Paris: Les Empêcheurs de Penser en Rond.

Lippmann, W. (1922). *Public Opinion*. Sioux Falls, SD: Nuvision Publications.

Marres, N. (2005). *No Issue, No Public: Democratic Deficits after the Displacement of Politics*. Ph.D. diss., Universiteit van Amsterdam.

———. (2007). "The Issues Deserve More Credit: Pragmatist Contributions to the Study of Public Involvement in Controversy." *Social Studies of Science* 37 (5), 759–780.

Marres, N. & Rogers. R. (2005). "Recipe for Tracing the Fate of Issues and Their Publics on the Web." In B. Latour & P. Weibel (eds.), *Making Things Public* (pp. 922–935). Cambridge, MA: MIT Press.

McKelvey, F. (2008). "The Software Politics of Web 2.0 in/through Drupal." Paper presented at Politics: Web 2.0: An International Conference, Royal Holloway, UK.

Mejias, U. (2008). "Networks and the Politics of the Para-Nodal." Paper presented at Politics: Web 2.0: An International Conference, Royal Holloway, UK.

Nelson, M. (2007). "MMP Goes Down to Defeat." Available at http://www.thestar.com/Ontario%20Election/article/265607.

Petersen, S. M. (2008). "Loser Generated Content: From Participation to Exploitation." *First Monday* 13 (3). Available at http://www.uic.edu/htbin/cgiwrap/bin/ojs/index.php/fm/article/view/2141/1948.

"Results Are In: Polls Are Dicey." (2007). *Montreal Gazette*. August 5. Editorial/Op-Ed.

Rogers, R. (2006). *Information Politics on the Web*. Cambridge, MA: MIT Press.

Scholz, T. (2008). "Market Ideology and the Myths of Web 2.0." *First Monday* 13 (3). Available at http://www.uic.edu/htbin/cgiwrap/bin/ojs/index.php/fm/article/view/2138/1945.

Szklarski, C. (2008). "CBC 'Wish List' Experiment with Facebook Dogged by Controversy." *The Globe and Mail*, June 28. Available at http://www.theglobeandmail.com/news/technology/article742008.ece.

Zimmer, M. (2008). "The Externalities of Search 2.0: The Emerging Privacy Threats When the Drive for the Perfect Search Engine Meets Web 2.0." *First Monday* 13 (3). Available at http://www.uic.edu/htbin/cgiwrap/bin/ojs/index.php/fm/article/view/2136/1944.

CHAPTER 4

Google Votes Australia: Portals, Platforms, and Embeds

On September 13, 2007 Peter Garrett, the Australian Labour Party's shadow minister for the environment and former front man for the popular '80s pop group Midnight Oil, walked on stage and proclaimed to a room full of reporters, "this will be a Google election."[1] Flanked by Liberal Party MP Joe Hockey, Garrett put aside his partisan talking points for an hour—just weeks before the beginning of the federal election campaign—to wholeheartedly endorse an ambitious online election campaign aggregator: Google's "Australia Votes" portal.

Google's Australian election portal was not only endorsed by political candidates, it also served as a quasi-official election site since all six of Australia's political parties agreed to provide unique campaign-related content. The portal thus created both a "monopolistic" online space—the only website with exclusive election-related content produced by all the political parties—with emergent Web 2.0 conventions that required the active uploading, sharing, and remixing of so-called user-generated content. The Google Australia Votes project thus serves as an important historical case study for subsequent forms of political campaigning, as it highlights how the world's most important networked information aggregator—Google's search engine portal—integrated its newly acquired video aggregator[2]—the YouTube platform—to produce an embedded network of political information. This new architecture of info-politics allowed individuals to host videos and other third-party content on their respective websites and blogs, while at the same time generating more traffic to the host social-media Web platform. In spite of Google's initial insistence that YouTube would "operate independently to preserve its successful brand and passionate community"[3] when it bought the site in October 2006 for the staggering amount of USD$1.65 billion, the Australia Votes project served as one of the first examples of the two Internet giants collaborating to produce a new form of networked politics, one harnessing a distributed network of users, digital objects, and multiple Web platforms. While previous chapters examined online politics by focusing on political actors outside of the social media platform model (bloggers), and on the ways in which other categories of actors such as issue-publics invested in and were defined by social networking platforms such as Facebook, this chapter focuses on the management of a political campaign via a networked social media platform, one characterized by a logic of embedded decentralization. In relation to the actor-object-platform triangulation developed in the introductory chapter of this book, the present chapter focuses on how the platform manages the relationships between political actors and their objects, that is, between the party apparatus and the official video campaigns within a distributed information model.

This chapter offers a case study of Google's Australia Votes platform for both technological and political reasons. First, it is an attempt to understand the shifting terrain of networked politics upon which permanent campaigning is enacted. This project was an important turning point in the history of networked politics and the Internet; one that witnessed a new architecture of participation for online partisans, bloggers, citizens, reporters, parties, and their candidates. This case study in particular highlights a key moment of transition in networked computing: the demise of Web portalization (one-stop shopping sites that provided access to the Net and personalized online content) and the subsequent rise of distributed embedded content and "host" Web 2.0 platforms (such as YouTube and Facebook) untethered from unique starting points or hubs on the Net. This new networked architecture calls for a method of analysis that seeks to determine the relationship between online context (digital objects), online users (those who upload, comment on, or "remix" content), and the Web platforms upon which networked politics is enacted.

Moreover, by focusing on Web 2.0 platforms, this chapter offers an empirical study of online political videos immediately prior to and during the Australian election in the fall of 2007. Our study of campaign videos specifically seeks to enumerate and track the multiplatform distribution, embedding, and viewership of campaign-related videos in an effort to understand the new socially mediated geography of networked politics—where videos are watched (embedded in various Web formats and sites) and where they are hosted (the Web platform that facilitates their storage and governs their ability to circulate across the Web and beyond). In the new 2.0 political landscape, the Google Australia Votes project highlights how media portals and platforms ("hosts") work in tandem to mediatize (Couldry, 2008; Hjarvard, 2008) the electoral process; that is, how they work together to make political campaigning "media friendly," or in our study, *to format politics for the social Web*. Our study of the Liberal and Labour Parties' YouTube-hosted videos thus seeks to understand how researchers of online politics can analyze and evaluate the conduct of political life in a mediated environment characterized by a distributed, cross-platform network architecture that governs the mobility of media objects, the visibility of political actors, and the connective capacities of new communication and campaigning applications.

The focus on YouTube-hosted videos therefore is not offered as a platform-specific study, nor even a medium-specific one (e.g., online videos). Rather, in addition to more broadly seeking to understand the demise of the Web-portal model of information aggregation, access, and personalization, our focus on YouTube campaign videos and their hosts seeks to determine how Google's Australia Votes project mediatized political debate during the campaign period and how it provided a new communicative framework for political debate among the key candidates. This chapter analyzes the network of YouTube videos produced by the leading Australian parties in 2007 to determine what other Web 2.0 platforms, campaign sites, or mainstream media outlets hosted the project's campaign videos. This study thus seeks to map the emergent platform politics during the campaign: the sites where candidates, activists, supporters, and campaign reporters embedded YouTube videos.

By identifying the source or host site of these embedded videos, this chapter provides insight into how distinct campaigning modes—in this case, the week prior to the start of the official election campaign and the last four days of the campaign, ending on voting day, November 24, 2007—determine what sites most actively contribute to the viewership of election videos. In an effort to distinguish between modes of permanent campaigning, our two studies offer a view of the dissemination of online politics beyond video aggregators, hosts, and platforms to the range of Web 2.0, mainstream media, and campaign websites that engage in various forms of electoral work—such as campaigning, debates, analysis, and reportage.

Google Australia Votes: The Portalization of Politics

After receiving endorsements and participation agreements from all of Australia's top political parties, Google's communications and marketing office in Australia spent the weeks leading up to the official start of the federal election campaign promoting its plans for the election to media outlets and citizens around the country. Google Australia Votes was characterized as a key site for user-generated content, an "online hub," and "Australian election central."[4] The portal itself contained four distinct yet complementary functions, all of which drew upon Google's search algorithm, information archives, and new data visualization tools. The user-generated online video archive YouTube stood at center stage.

Each of Australia's six political parties agreed to upload their official videos to Google Australia Votes–branded "channels," a new format for organizing YouTube videos that sought to play upon the traditional formatting of television broadcasting and programming while also producing a new personalized space for individual users to design and host their video archive, viewer lists, and favorite friends (see Figure 1). The new YouTube "channel" format also dovetailed with YouTube's decision to repost its original trademarked motto "Broadcast Yourself" next to its corporate logo. Google also hosted their own "Australia Votes" channel on YouTube for video commentary on the project and the campaign in general.

Figure 1. Branded YouTube Channels for Google's Australia Votes Portal. Courtesy of Google Inc.

Next to YouTube video channels, the portal's most prominent features drew upon Google's geographic visualization tools, specifically the Google Maps and Google Earth interfaces that serve as platforms for an infinite number of data sets. For the Australia Votes project, Google Maps predictably started with electoral boundaries (see Figure 2) that henceforth served to highlight through color coding, search boxes, and pop-up windows information such as potential swing ridings (that typically switch parties in elections) and information on the candidates (their websites, voting records, etc.). Google searches for videos, news, and other documents were also integrated into almost every level of such geographical interfaces.

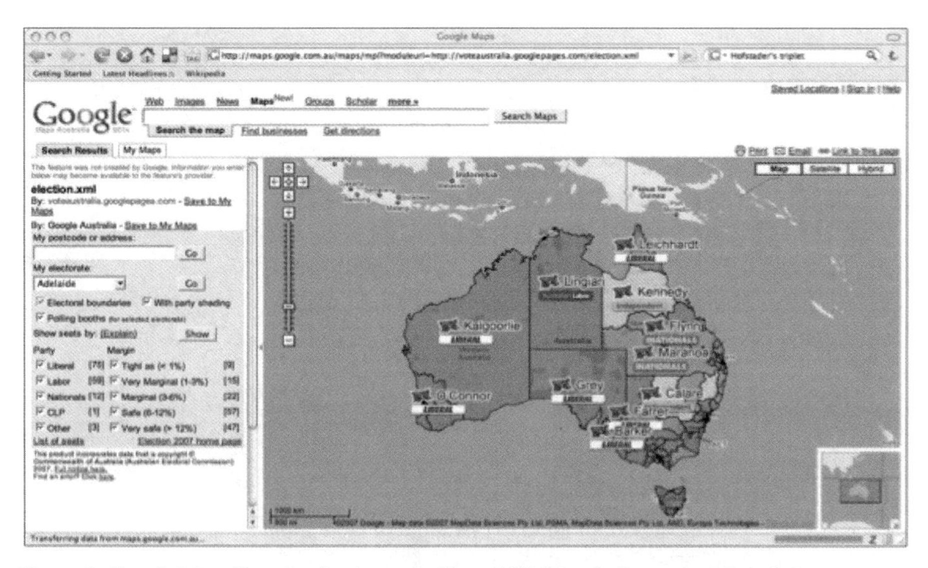

Figure 2. Google Maps Function for Australia Votes 2007 Portal. Courtesy of Google Inc.

Lastly, mirroring a general trend within Google's properties in 2007, the Australia Votes portal highlighted a number of personalized search engine gadgets. The suite of tools integrated into Google's personalized search engine page, dubbed "iGoogle," was first tested and promoted at a Google Personalization Workshop held at its Mountain View campus in April 2007. iGoogle was developed to expand Google's search functions offline to include the PC desktop space, to integrate personal search histories into search results, and for the first time, to "push" recommended services and content to individual users. iGoogle was to be powered by add-on "gadgets," small software tools that build upon the new personalized information architecture to deliver more localized and personalized services. By the spring of 2007, iGoogle offered more than 25,000 such gadgets.[5] This new search engine architecture consequently provided opportunities for programmers to contribute to Google's suite of services, a point that was emphasized during the launch of the Australia Votes project.

Four unique gadgets were designed by Google specifically for the Australia Votes site: (1) a "news from your seat" gadget (see Figure 3) that pushed election news to the election riding of the user; (2) a YouTube gadget that listed the latest uploads from all Australia Votes video channels; (3) an "on the record" gadget that users could use to search for a politician's views on specific social issues; and lastly, (4) a trends gadget (a precursor to Google's popular analytics service) that tracked the frequency of both election news reports and search engine search terms ("keywords").

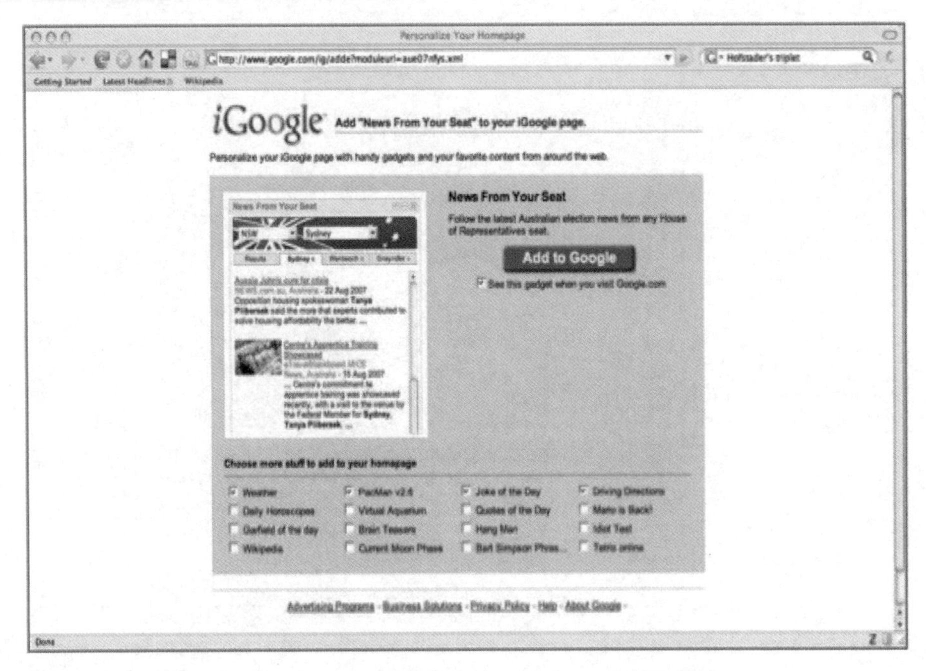

Figure 3. The "News From Your Seat" iGoogle Gadget. Courtesy of Google Inc.

The Web 2.0 Election: Analyses of the Australian Campaign

Given Google's promotional apparatus, its compelling set of gadgets, maps, and functions, and most importantly, its widespread adoption by and participation from federal political parties and candidates, one would expect the Google Australia Votes portal to be a key aspect of some Internet campaign studies. Curiously, however, scholars of the 2007 campaign—even those who focused exclusively on the role of new media and the Internet—almost entirely ignored the landmark project. Macnamara and Bell's (2008) comprehensive 69-page study of "e-electioneering" in the 2007 election, for example, quotes Peter Garrett's proclamation of a "Google election" (p. 6), but subsequently makes no mention of the Google portal. They choose instead to focus their attention largely on the role of YouTube. The media's preference for branding the campaign through the YouTube platform rather than its corporate owner and Web portal can perhaps be explained in part by the broader publicity surrounding a series of campaign-style ads posted on YouTube by the sitting prime minister, John Howard, and Labour's Kevin Rudd in the months preceding the election campaign.

A staff paper from the National Library of Australia provides one of the very few reports that explicitly elaborates upon the Google project, which next to the Australian Broadcasting Corporation's Web-based coverage of the election produced the largest archive of political content (5 gigabytes in size), and received more than 700 links from online videos (Crook, 2007, p. 3). The National Library report also notes that the project's YouTube component, having started with six channels (the aforementioned project channel and those of the political parties), grew to sixty-four channels by the end of the election campaign (p. 4).

Howell and Da Silva's (2010) study of the effectiveness of Web 2.0 technologies and platforms in engaging first-time voters in Australia is much more representative of the new media literature on the election. The authors claim that new media tactics had two goals in the election: "First, to attract youthful voters and second, to educate and deliver policy on a level generally associated with the 18–24 demographic" (p. 28). While the authors impart a common "YouTube election" mantra repeated throughout Australia's mainstream media, they too omit any mention of the Google project. What makes Howell and Da Silva's omission of Google's Australia Votes election portal particularly interesting is their stated goal to map the relationship between new social media platforms in an effort to determine the flow and circulation of content and locate the source of election information for young and first-time voters. Howell and Da Silva locate YouTube, and not Google, as the key hub or mediator of election material among first-time voters, party websites, and social networking sites (p. 29).[6] (See Figure 4.)

By ignoring the Google portal and Australia Votes project in general, Howell and Da Silva fail to recognize Google's key role in what Chen (2010) calls—in his comparative study of the 2007 Australian election with Canada and New Zealand—the new "media ecosystem." In other words, Howell and Da Silva do not take into consideration the structural link between the political parties and the YouTube platform—the fact that all parties had agreed with YouTube's parent company to post their official videos on the Google Australia Votes–branded YouTube channels. Moreover, such an omission led the authors to conclude that video and social networking sites only serve to produce "online interpretations… from a singular source (party websites)."[7]

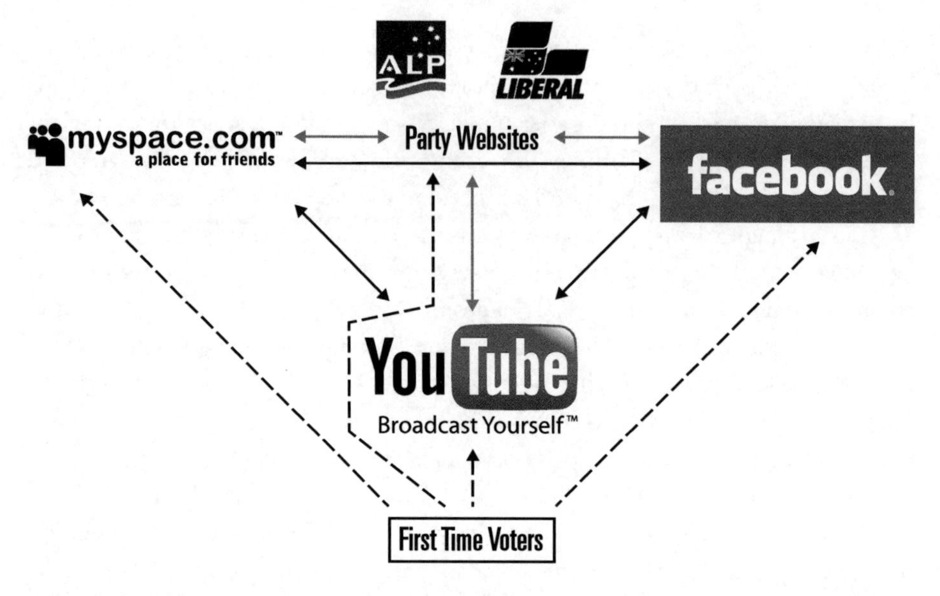

Figure 4. Taken from Howell and Da Silva's (2010) "New Media, First Time Voters and the 2007 Australian Federal Election" with permission from the authors.

While Howell and Da Silva conclude that YouTube and social networking sites merely served to *mediate* political content from the political parties to first-time voters, this chapter argues that the Google portal served to mediatize the Australia electoral process. Nick Couldry (2008) contends that mediatization offers a much more expansive and structural concept than "mediation," providing a lens through which we can understand how various cultural and political processes, as well as conventions, are prepared for the exigencies of media production and consumption. In Couldry's words, "many cultural and social processes are now constrained to take on a form suitable for media representation" (p. 376). The Australia Votes platform, in conjunction with a new social media landscape that encouraged individuals to upload, share, remix, and comment upon digital media and text, effectively mediatized the political process. In short, this new networked politics set the stage—through an expansion of the sites, spaces, and platforms of political networking, communications, and campaigns—to require an ongoing, permanent form of political campaigning.

The Google Australia Votes project is a transitionary moment on—and for—the Web. Its processes involved the reportalization of content and services from prominent info and default-set hubs on the Web—such as Google—to embedded networks that include individual websites, blogs, and social media platforms. This transition to a Web 2.0 architecture in effect made mobile the digital objects of political communication (videos, images, blog posts, comments, reviews, news pages, etc.), while at the same time maintaining connections to—or better, hosted by—their parent platforms such as YouTube. Such host platforms, in the first instance, enabled the uploading, hosting, and most importantly, the sharing of digital objects on the Internet.

Such a mediatized interpretation of Google Australia Votes was clearly in evidence at the project's initial press conference, where representatives of Google attempted to wrap the project into an all-encompassing democratic ethos wherein the project represented "the convergence between democracy and the Internet."[8] Moreover, as an effort to mediatize the conventions of political communication, reporting, and campaigning, Google did not direct its project solely to the lay user—that is, the voter. Rather, as Google's executive noted at the launch of the portal, Australia Votes would serve as a "powerful starting point for research for both voters and journalists, and... candidates as well."[9] Days after the Google press conference with Garrett and Hockey, Google held a lunch for federal MPs to further discuss how the site might help their respective campaigns.[10] Moreover, in addition to pulling in political participants and pushing out personalized and aggregated content, Google's portal also sought to mediatize the new networked landscape, situating itself—principally through its YouTube property (and platform)—at a new "center" of the Web 2.0.

From Portals to Platforms

As an innovative leader in data aggregation and network services, Google's Australia Votes project was not the first effort to portalize content and services for users (or in this instance, Australian voters with Internet access). In an effort to create common sites from which to start browsing the Internet and the Web, portals have long sought to structure users' experience of networked communications. Korinna Patelis (2000) defines a portal as a force of functionality and customization; that is, the portal is useful to the user because it orders the chaos of the Web by bringing together information important to the user (p. 50). While Google has long sought to actively integrate users, producers, and programmers of information into its business model, previous online portals were largely modeled along the lines defined by Patelis. These portals operated primarily as customer service formats—interfaces that enabled network access, management of resources and communications (most notably email), and to varying degrees, the personalization of Web-based content.

Portals also predate the Internet as a means of accessing digital networks. Personal computer hobbyists initially used telephone modems to log onto electronic bulletin boards hosted by individual servers to engage in a host of networking activities (Murphy, 2002). Commercial portals soon followed, offering their customers a unified experience, easy access, and aggregated content. There would be no need to sift through "the Internet," and moreover, one's attention would be gracefully managed through the adoption of newspaper-like sections: sports, news, weather, and so forth. Such portals—dominated by the likes of AOL, Compuserve, and Prodigy in the U.S.—were in retrospect spectacularly complacent or passive; they treated their users as consumers and not as producers of content.

While leaving Internet access to Internet service providers, Google has over the years sought to establish their search engine as a key hub or portal to start browsing and searching for Internet-based content. Google's purchase of YouTube served notice that the portal model of networked communications was nearing its end, or at least was undergoing a fundamental change. It proved there were new tools, algorithms, and spaces for the uploading, manipulation, and circulation of user-generated content. These new platforms—YouTube, Myspace, and most prominently Facebook—all developed forms of sharing content through personal accounts that were encouraged to link to friends, colleagues, and other like-minded users. Each of these new platforms developed its own distinct sets of practices, conventions, and governing algorithms that "sat" on top of pre-existing Web protocols. Such sites, however, soon realized that while their own properties needed additional users and expanded social networks to generate advertising revenue, they would also need to develop cross-platform protocols for the sharing, and later embedding, of videos, text, images, and other digital files. YouTube was subsequently among the first social media platforms to recognize that their site could provide unique tools, services, and social networks to host users, viewers, and producers while expanding their network presence by making all of their videos mobile—meaning that the code for YouTube-hosted videos could be transported and embedded in most other Web formats, including blogs, mainstream media sites, and other social networking sites.

Embedded Methods: The Hosting of Platform Politics

Our search to determine how the new embedded network of political campaigning mediatized the 2007 Australian election began by constructing a small yet significant sample of political videos produced and uploaded by the Liberal and Labour Parties, the two leading parties in the 2007 election. The study used YouTube's statistics to determine the top five most-viewed videos for Kevin Rudd (Labour) and John Howard (Liberal) before and during the campaign. The top five videos were collected on October 17, 2007, the same day the writ of election was dropped. Since most of the top videos had been collected in the late summer, these videos represent their total viewership until the day we visited the site. The second sample ran on November 20, 2007, four days before the election, with the intention of determining the most popular videos during the campaign. Again, we collected the videos that attracted the most views.

With the top-viewed videos for both leadership accounts in hand, we set out to understand how the Google project—and in particular its YouTube platform—mediatized the conventions of electoral communications and campaigning. In an effort to better understand, in comparative terms, the forms of campaigning enacted on and across the Web, we decided to study two distinct time periods. In particular, we were concerned with how the Google project served to format certain possibilities or platforms for campaign events, dialogues, and debates among the parties, leaders, candidates, and a host of other political actors, activists, and commentators.

As previously noted, the Google project was not the first instance of YouTube-enabled political jostling in Australia. Throughout the summer of 2007, the Labour Party—who at the time were leading in the polls—and the incumbent Liberal Party exchanged a set of YouTube-hosted attack ads. As the following charts demonstrate (see Figures 5 and 6), the Liberal Party, under the helm of its unpopular prime minister, John Howard, was struggling to find viewers for its videos, and more broadly, for its policies. Key Liberal Party insiders and traditionally supportive newspaper columnists had begun to openly advocate for a new party leader, just months in advance of the fall election (Bongiorno, 2008, p. 595). Labour's Kevin Rudd, the subsequent winner of the fall 2007 election, succeeded in attracting over 100,000 views for just one of his party's YouTube videos. The video demonstrated a populist-style appeal that encouraged Australians to identify their country's most pressing social, political, and economic issues ("Tell Kevin Rudd: What Issues Are You Passionate About?"). Even before the campaign had officially been called, voters were flocking to a video that—rhetorically at least—sought to take advantage of the interactive possibilities of social media. By the start of the official campaign in October 2007, Labour had posted fifty-seven videos on their YouTube account. The governing Liberals, by comparison, had posted only thirty-four.

Title	Date Uploaded	Views
Tell Kevin Rudd: What Issues Are You Passionate About?	6 August 2007	100,813
Who Will Howard Blame for the Next Interest Rate Rise?	5 August 2007	35,616
John Howard Asleep on Climate Change	16 July 2007	27,487
Confused? So Are John Howard and Peter Costello!	12 September 2007	16,417
John Howard Promised to Keep Interest Rates at 'Record Lows'	8 August 2007	13,566

Figure 5. Top Five Most-Viewed Kevin Rudd (Labour) Pre-Campaign Videos.

Title	Date Uploaded	Views
Labour Can't Manage Money	5 August 2007	13,516
Prime Minister John Howard	Unknown	12,029
Prime Minister's Well-Being Plan for Australian Children	21 September 2007	4,687
Prime Minister John Howard's Defence Gap Year Announcement	8 August 2007	1,061
Prime Minister's Mersey Hospital Announcement	31 July 2007	975

Figure 6. Top Five Most-Viewed John Howard (Liberal) Pre-Campaign Videos.

Viewership numbers, and the number of videos produced and uploaded by parties, provide a helpful yet limited understanding of the "Google election." More importantly, from this pre-campaign picture we can see that Labour had already indicated, through its vastly popular video in the summer of 2007, a shifting of gears in its communications and campaigning strategy. In other words, Labour sought to adopt the ethos of the social media functions of the YouTube platform to critique the governing Liberal leader while at the same time experimenting with the new language of socially mediated politics—a language of feedback and participation. John Howard's pre-campaign videos, by comparison, often framed the Liberal leader and incumbent prime minister as directly addressing YouTube viewers. Unlike Rudd, Howard's videos lacked any rhetoric or appeal for online participation. Coupled with his other most-viewed videos, typically announcing daily events in advance of a "real world" photo-op or press conference, these videos indicated a largely unidirectional voice from the prime minister to voters.

The Campaign

The election writ dropped on October 17, 2007, setting the stage for five weeks of campaigning. It ended with voting on November 24. The campaign lasted precisely thirty-nine days, in a country where the minimum is thirty-three days and the maximum is sixty-eight days. By the conclusion of the campaign, Labour received 64,679 views for their top five videos, while there were 47,796 views of the Liberals' top five videos.

Title	Date Uploaded	Views
Another Day, Another Stale Scare Campaign From Mr. Howard	16 October 2007	18,881
Climate Change: You're Voting for the Future on Nov. 24	22 October 2007	15,512
Mr. Howard's Scare Campaign on the Economy Lacks Credibility	18 October 2007	11,313
A Plan for Australia's Future—Not a Stale Scare Campaign!	15 October 2007	9,761
Really Mr. Howard?	3 November 2007	9,212

Figure 7. Top Five Most-Viewed Kevin Rudd (Labour) Campaign Videos.

Title	Date Uploaded	Views
It's Scrutiny, Not a Scare Campaign	17 October 2007	21,663
Prime Minister John Howard's Election YouTube	14 October 2007	8,535
Labour Lets the Cat Out of the Bag	4 November 2007	7,321
Prime Minister John Howard's Skills Vouchers Announcement	28 October 2007	5,223
Who Will Really Pull the Strings?	28 October 2007	5,054

Figure 8. Top Five Most-Viewed John Howard (Liberal) Campaign Videos.

A major theme of the top five Labour campaign videos is a unique and arguably postmodern video thread style, a novel format that went largely unnoticed by scholars of the campaign. Three of the most-viewed videos featured this format. These videos begin with a television showing a Howard clip and then zoom out of frame to show Rudd holding a remote control. Rudd subsequently stops the clip and goes on to refute scare tactics depicted on the Liberal's election ad. The video-within-a-video format did not end there, however. The Liberal campaign responded in kind, producing an Internet-like threaded conversation or debate. The most popular campaign video for the Howard campaign begins with a TV playing a Howard clip, then zooms out to show Rudd refuting the original, and then zooms out yet again to show Howard holding the remote control, stopping the Rudd clip and refuting its claims while sitting at a desk. What such a threaded debate indicates is that Google's project, and participation by the parties in this project, produced a new space—perhaps we could call it a platform—that opponents could not only seize upon in very short order, but also very easily mimic in style and refute in content. The bringing together of the parties under one roof, so to speak, putting the possibilities for interactive dialogue and debate on one site, mediatized the election. It called for the parties to imagine new ways to speak to the other side and to undercut the possible advantage that the competition might have with their arrangements and relationships with other media outlets and broadcasters.

To suggest that either Google—or more to the point, YouTube—portalized the campaign, however, would neglect the importance of where users viewed these threaded videos: the new embedded spaces where campaign videos were promoted and accompanied or juxtaposed to other content across the Web. For the most part, views were generated from the embedding of videos across the Web. This is because YouTube acts as a video database—an archive of online video content—designed to be embedded in other media. Users often refer to the site not as a space of discussion and interaction, but as a resource to host the videos that they refer to elsewhere. Bloggers, for instance, host videos on YouTube, but post them on their blogs as embedded videos. Viewers, then, have little interaction with the site other than the small buttons around the video embedded on the blogs they visit.

To determine the dissemination and embedded reach—or network—of videos from the Liberal and Labour Parties, we sought to determine the sources of views from their host embeds. When the research was conducted for this chapter—prior to YouTube developing an application programming interface (API) where programmers and others could access such data—the platform offered only a breakdown of sites where views were registered (see Figure 9). Total links for the two periods under study were then charted to determine what types of sites promoted the most embedded views of the Labour and Liberal Parties' uploaded videos.

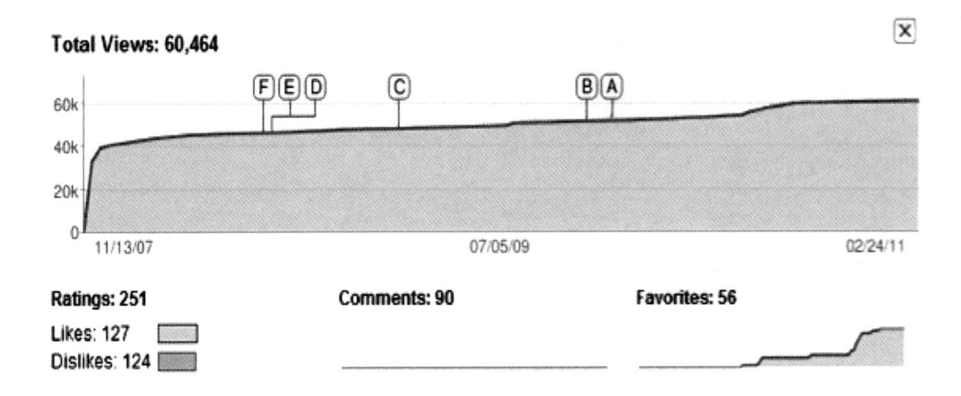

Links

	Date	Link	Views
A	12/ 1/09	First referral from related video - Taking Australia Backwards	813
B	10/27/09	First view from a mobile device	295
C	01/31/09	First embedded on - blogs.news.com.au	1,135
D	08/13/08	First embedded on - encyclopediadramatica.com	2,070
E	08/12/08	First embedded on - www.encyclopediadramatica.com	440
F	07/25/08	First referral from - www.facebook.com	767
	Unavailable*	First referral from related video - The Chaser's War on Everything: Kevin Rudd Labor Commercial	2,710
	Unavailable*	First embedded view	631
	Unavailable*	First referral from related video - Really Mr Howard?	339
	Unavailable*	First referral from YouTube - Homepage	313

*Exact date is not available. The count of views from this link only includes views since 30 November 2007.

Audiences

This video is most popular with:

Gender	Age
Male	45-54
Male	35-44
Male	18-24

This video is most popular in:

Honors for this video (0)

(There are no honors for this video.)

Figure 9. YouTube Video Statistics. Courtesy of Google Inc.

The study coded each link by its platform type, such as a mainstream media website or a social networking blog. Coding the links by platform type provides a glimpse of not only where political videos were hosted, but also where they were viewed, since they contributed to the most-viewed videos of the two campaign periods.

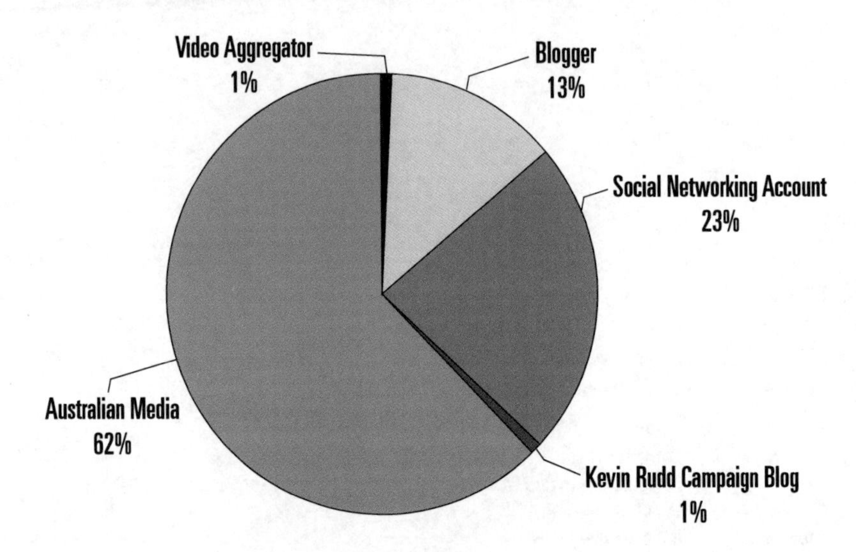

Figure 10. Top Referrals to Official Labour Party YouTube Channel Videos (Pre-Campaign).

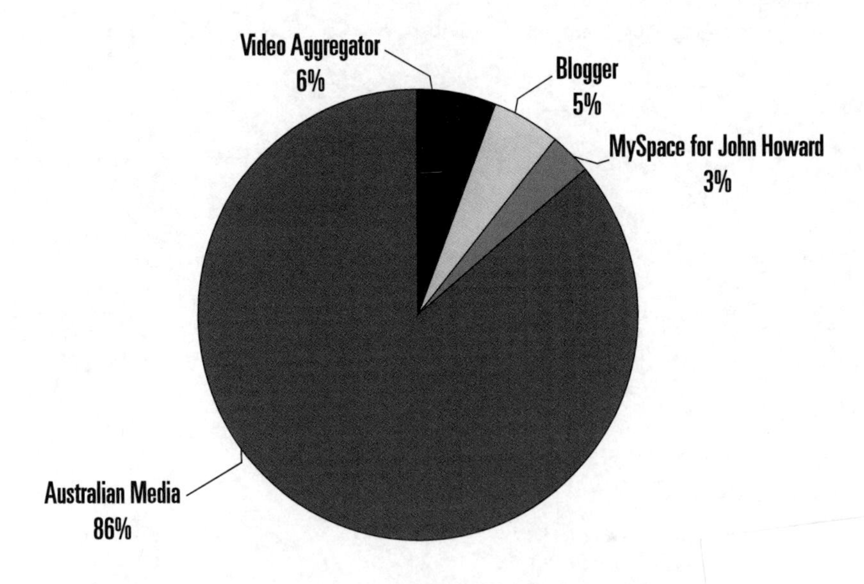

Figure 11. Top Referrals to Official Liberal Party YouTube Channel Videos (Pre

During the pre-campaign period there was a marked difference in the sources of viewership between the two leaders' top-viewed YouTube-hosted videos. Although the majority of links to the Rudd videos came from the Australian media (1,617 links), there was a greater plurality of referral types than for Howard's videos. Bloggers provided the embedded views for many more of the Rudd pre-campaign videos (330 links from Blogger) than the Howard videos (64 links from Blogger), suggesting that the Howard campaign lagged behind Rudd's in having their videos distributed, debated on, and discussed across the blogosphere. In this pre-campaign period, Rudd's videos were also more widely embedded on social networking pages such as Facebook and Myspace, producing 604 links in total.

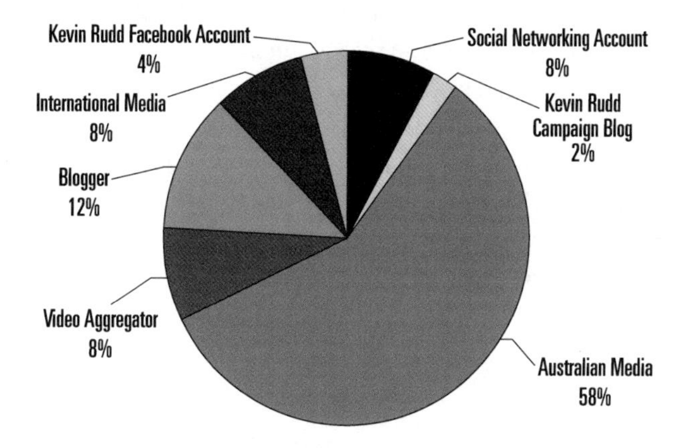

Figure 12. Top Referrals to Official Labour Party YouTube Channel Videos (During the Campaign).

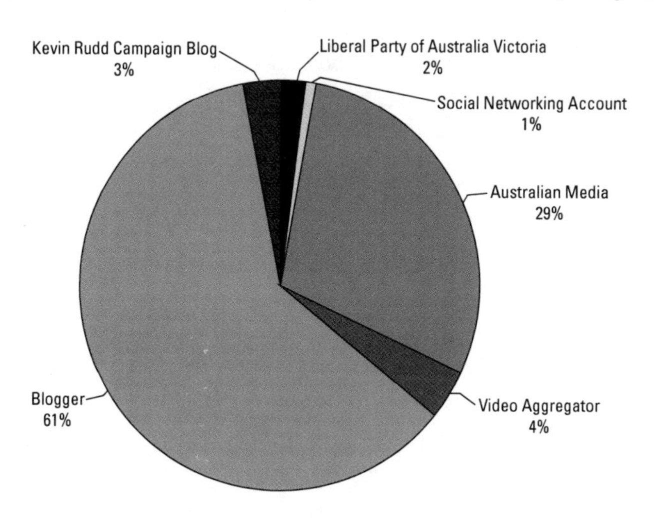

Figure 13. Top Referrals to Official Liberal Party YouTube Channel Videos (During the Campaign).

During the campaign (see Figures 12 and 13), the Howard (Liberal) campaign received dramatically more links from bloggers (713 links) compared to the pre-campaign period. Most of these embedded views were attributable to an Australian anti-Rudd blogger "andrewlanderyou" who posted more videos with the commencement of the election campaign. Embedded views from blogs for Rudd, however, showed little change from the pre-campaign to the campaign period (12 to 13 percent as a total of all views). While Howard's campaign seemed to get a boost from the banter of the blogosphere, Australia's mainstream media were almost twice as likely to produce embedded views of Rudd's videos (658 links) than those of Howard (341 links). International media continued this trend, producing eighty-eight views of Rudd videos (88 links). Labour, unlike the Liberals, seemed to capitalize on social networking and video aggregators as a means of referring Internet users to campaign videos. Whereas Howard received embedded views from only a few platforms (blogs and Australian media websites), Rudd continually received embedded views from a variety of platforms (his own social media accounts, social networking accounts, and video aggregators).

Conclusions

The source or location of embedded videos thus demonstrates that the ability to make the objects of political campaigns mobile (in the instance of online videos, via their embedded code) greatly extended the reach of the Google project beyond its search engine, the Australia Votes portal, and the YouTube platform itself. Rudd's video network suggests that his campaign sought to integrate his own blog and social networking sites into the embedded economy of YouTube and Google's Australia Votes, whereas Howard's embedded viewership suggests that he was more widely discussed and debated by partisan bloggers, an arena that lends itself to more substantive policy analysis and partisan debating.

From a networked perspective, we can conclude from our analysis of embedded videos that the different campaigns drew attention from different sources, or rather from different platforms that have their own set of characteristics, affordances, and roles in political campaigns. The ability of Google to "portalize" the election can thus be understood to be a participatory move on its own. Bringing official political actors together under one roof, as it were, greatly heightened Google's role in mediatizing the campaign—and in forcing political actors to conform to the exigencies of Google's business model. Such a role was clearly at play, not only in portalizing the campaign's official electoral actors (the parties and candidates), but also in deploying an embedded economy of platforms.

One of the key goals of this chapter was to highlight how Google's Australia Votes project served as a key moment in Internet history and as a transitional moment in the reconfiguration of networked portals. Google continued to experiment with push technology, geolocative tools, and interfaces, tweaking its search algorithm and many other technologies and assets during this period. However, the deployment of their newly acquired YouTube platform at the center of the election campaign suggests that they recognized the possibilities of the emerging network architecture as an increasingly distributed network that would serve to enable users to upload, download, and modify mobile media objects. This project presents an opportunity to uncover and map the conventions, platforms, and traffic of an emerging user-generated politics; one that challenged the parties to develop strategies for speaking to and across an embedded network of media objects and newfound hosts. Such a mediatized political process poses unique challenges for political parties. Central to this is the need to manage, or rather steer, the spread and embedding of content to particular supporters. Google's Australia Votes project demonstrates the need for political actors, candidates, and parties to engage in a parallel battle over the mastery of the portalized sphere of official politics, and at the same time, over the rapidly changing rules and conventions of platform politics.

Indeed, one of the key issues emerging from this chapter is the recognition that the Web 2.0 platform model of participation, sharing, and remixing not only fosters processes of decentralization of politics whereby the objects (e.g., campaign videos) follow distribution paths beyond traditional channels, but also creates new pressures for political actors to manage such new modes of distribution and circulation of content. The logic of the network portal model expresses this paradoxical relationship quite strikingly: The network portal is not only about accommodating an endless supply of decentralized, user-generated, participatory content, it is also about managing this content with regards to making it accessible and visible to users. Web 2.0 platforms in general, as we have seen in this chapter and the previous one, develop their own particular algorithmic logics for organizing decentralized information. The challenge, however, is that such logics of constantly ordering information and content provoke an acceleration of political communication: Repeated slogans, political discourses, and other discursive paraphernalia of political party communication have to be confronted with the potential decentralization and hijacking of its messages through practices of remixing, re-uploading, and commenting. As such, time, and more particularly, the acceleration of the tempo of political communication, becomes a new strategic site of the online permanent campaign, as we see in the next chapter.

Notes

1. The entire Google Australia Votes press conference can be viewed on YouTube at http://www.youtube.com/watch?v=s3Q61a0epIM.
2. Google announced on October 9, 2006—approximately a year before the announcement of the Google Australia Votes portal—that it had acquired the online video-sharing site YouTube for USD$1.65 billion in stock. See http://www.google.com/press/pressrel/google_youtube.html.
3. Available at http://www.google.com/press/pressrel/google_youtube.html.
4. Available at http://www.youtube.com/watch?v=s3Q61a0epIM.
5. Available at http://blogoscoped.com/archive/2007-04-30-n90.html.
6. The social networking sites featured in their study are most notably Facebook and Myspace.
7. See Howell & Da Silva (2010), p. 29.
8. Ibid.
9. Available at http://www.youtube.com/watch?v=s3Q61a0epIM.
10. Available at http://www.google.com.au/press/pressrel/election07.html.

References

Bongiorno, F. (2008). "Howard's End: The 2007 Australian Election." *The Round Table* 97 (397), 589–603.

Chen, P. J. (2010). "Adoption and Use of Digital Media in Election Campaigns: Australia, Canada and New Zealand." *Public Communication Review* 1, 3–26.

Couldry, N. (2008). "Mediatization or Mediation? Alternative Understandings of the Emergent Space of Digital Storytelling." *New Media & Society* 10 (3), 373–391.

Crook, E. (2007). The 2007 Australian Federal Election on the Internet. Canberra: National Library of Australia Staff Paper (pp. 2–17). Available at www.nla.gov.au/openpublish/index.php/nlasp/article/.../1040/1308.

Howell, G. & Da Silva, B. (2010). "New Media, First-time Voters and the 2007 Australian Federal Election." *Public Communication Review* 1, 3–26.

Hjarvard, S. (2008) "The Mediatization of Society: A Theory of the Media as Agents of Social and Cultural Change." *Nordicom Review* 29 (2), 105–134.

Macnamara, J. & Bell, P. (2008). "E-Electioneering: Use of New Media During the 2007 Australian Federal Election." Research Report. Australian Centre for Public Communication.

Murphy, B. (2002). "A Critical History of the Internet." In G. Elmer (ed.), *Critical Perspectives on the Internet* (pp. 27–48). Boulder, CO: Rowman & Littlefield.

Patelis, K. (2000). "E-Mediation by America On-line." In R. Rogers (ed.), *Preferred Placement: Knowledge Politics on the Web* (pp. 49–63). Maastricht: Jan van Eyck Editions.

CHAPTER 5

Live Research: Twittering an Election Debate

Mediated life has so vastly multiplied its forms and sites of communication and storytelling that the ability to recall where one heard or viewed a news report, a rumor about a friend, or even the source of an urgent work-related request now requires a panoply of aggregate technologies: smartphones, RSS feed managers, personalized search engines, live feeds on social networks, and so forth. The rapid growth of networked, handheld, virtual, embedded, and locative information and communication technologies raises important questions about methods of studying political processes, objects, actors, and technological platforms that are—by design or dysfunction—constantly in flux. In an age of meta-information, such technologies serve to collapse and focus time, which is increasingly socially mediated time, to a window of approximately 10 minutes. This occurs both in the past, through interfaces such as Facebook or Twitter that bury 10-minute-old communications, and in the future, through anticipatory buzzing and pinging reminders of duties to come in 10 minutes. Visually, such *interface time* literally hypermediates a window in time for that which can fit on the interface before being pushed off, or typically down, to make way for the next 10 minutes. How these new logics of accelerated time have transformed the objects of the permanent campaign is the focus of both this chapter and the subsequent one. This chapter and Chapter Six, by focusing on the transformation of the objects of the permanent campaign, also develop a methodological discussion: If, as we have posited, the ontology of political objects changes radically in the platform context, then what are the repercussions with regard to redefining traditional media research methods? What traditionally used to be understood as a fixed textual object (i.e., slogans, images, video) now has to be considered as always in circulation and always in the process of being rearticulated, both at the level of content (sharing, remixing, commenting) and at the level of the informational logic of the platform, from the me-centric to time-constrained logics. This chapter in particular focuses on the Twitter platform as a time-constraining platform, and examines the new political practices of object-creation that emerge in this particular context.

Unlike Facebook, Myspace, Cyworld, Bebo, and other social networking sites that offer a vast array of interfaces and functions for users and their networked friends, micro-blogging platforms such as Twitter offer a decidedly trimmed-down interface featuring a vertical ticker of short bursts of text (with a 140-character limit). Such an interface maintains a concise focus on a very brief window of time. Unlike horizontal stock or sports tickers that communicate incremental changes in prices and scores in a constant loop, Twitter's ticker relies upon friends and contacts to actively "retweet" a post back to the top of its vertical interface (visible again to the user). Initial research, however, has found that only 6 percent of all tweets are reposted to the top of Twitter's vertical interface.[1] Duncan Geere (2010) summarizes this point nicely: "92 percent of… retweets occur within the first hour. Multiplying those probabilities together means that fewer than one in 200 messages get retweeted after an hour's gone by. Essentially, once that hour's up, your message is ancient history."[2] Such findings question the means by which individuals or political campaigns might sustain and expand the readership of their posts across Twitter's social networks. Un-retweeted comments on Twitter resemble a hyperactive blog interface, whereupon newer posts push older ones down and, in short order, off a user's PC, tablet, or smartphone interface. Older posts are in effect buried into the interface-depths of the infinite downward scroll, or are pushed off onto additional hyperlinked pages via the indefinite "next page" click.

The emergence of vertical tickers on social media platforms and other forms of hyper-immediate, time-compressed social media interfaces highlights the need for real-time forms of Internet research. The "live" form of research we describe in this chapter seeks to investigate the techniques, technologies, and user dynamics that attempt to expand this intensely time-compressed platform during a live-broadcast political debate. This chapter demonstrates how political forms of communication—particularly during heightened periods of partisan conflict such as elections, scandals, and political or economic crises—are being expanded onto "second screens" (typically. PCs and smartphones running social media interfaces) that enable socially mediated conversations and networked commentary on live broadcast events. Live research requires an understanding of the networked affordances and technological encodings (e.g., meta-tags) of discrete digital bursts or *objects* (Schneider & Foot, 2010)—particularly tweets, blog posts, or comments posted on online newspapers' Web pages from their specific platforms, or from larger aggregators such as personalized feed (RSS) managers, search engines, or social networking sites. Such components consequently form the basis for software code-focused media research, the platform upon which researchers can attempt to determine the tactics, conventions, functions, and dysfunctions of real-time political discourse on social networking sites such as Twitter, or across mediated screens, platforms, and interfaces (Rogers, 2006).

Given the rapid development of social media platforms, user conventions, and the ever-changing sets of rules and regulations that govern sites such as Twitter (as manifest through their programming interface or API), our discussion of live research seeks to account for the always-already shifting dynamics in political communication flows, *both a permanent and immanent campaign.* While some may impart a Latourian (2005) motive at work, particularly with regard to his object-oriented philosophy (see also Harman, 2009), this study extends beyond the tweet-as-object to an appreciation of the temporalities of interfaces, information architectures, and the political tactics deployed on social media platforms. What is suggested herein is a more immediate and immersed form of research; that is, one that not merely "tracks the object" (Lash, 2007), but rather undertakes a reflexive, empirical approach to understanding media flows in social media's increasingly compressed interface time.

A focus on in-the-moment communications and networking attempts to build upon broader discussions, theories, and methods of critically engaging with open-ended networked, nonhierarchical, or distributed forms of communications (Fuller, 2003; Galloway, 2004) in order to understand the strategic deployment of political campaigns and communications in compressed and socially mediated interface time (Cunningham, 2008). The question of live research in Internet studies, and consequently in ICT-enabled studies of political communications (Chadwick & Howard, 2009; Kluver et al., 2007), continues to develop an important methodological debate within the broader field of Internet studies. Andrew Chadwick's (2011) recent study of shifting political information cycles are of particular importance to this form of live research. In attempting to determine the new roles and opportunities that social media afford in the political process, Chadwick investigates the temporality and flow of political news—building on Norris (2000) before him—so as to better understand how social media actors intervene and disrupt political and mainstream media tempos and schedules in real-time, in effect producing a new tempo of mediated political life, or in Chadwick's terms, a new "political information cycle."

Methods of real-time research, however, have a much longer history in Internet studies than this literature sampling suggests. Annette Markham's (1998) work on virtual chat rooms, for example, offered an auto-ethnographic approach to the study of computer-mediated communication as a distinctly participatory form of real-time or live research. Markham's study sought to enumerate the complex literacies involved in navigating a virtual chat room in the moment, for which she logged the challenges she faced onscreen as they occurred in real time. In this way, Markham's study highlighted important time-sensitive conventions that transpire in online environments, a process that was made all the more apparent by her recollections of being immersed in live interactions with other users, the software, and the interface itself. Christine Hine (2007) similarly suggested a "connective ethnographic" approach to understanding how various forms of computer-mediated communication connect users to their "offline" lives. It is this approach that informs the present work, as an attempt to understand how political campaigns and communications seek to reconnect political communication (images, blog posts, etc.) across social media interfaces. In so doing, we aim to redress the temporal limits of communication and subsequent limited attention span of new media audiences and social media interfaces.

Recent examples of live or real-time research have also emphasized the state of always being ready to conduct research; of being in a position to capture a political crisis or a live mediated event on the Net. Andreas Jungherr and Pascal Jürgens' study of Twitter use in Germany (2011), for instance, builds upon James Allan's (2002) notion of "topic detection" as an attempt to continuously collect and analyze social media content feeds and flows of information for signs of increased activity. While our project similarly utilizes a method of data collection and content analysis of tweets in advance of the televised election debate under study, we also seek to understand the tactical forms of political communication deployed in real time on Twitter and other Web platforms during the debate broadcast—strategic forms of communication advanced by political parties and other online partisan actors.

Overall, the live research paradigm places greater emphasis on the relationship between the rules and regulations of social media platforms as we move from a "news cycle" paradigm to one defined by the new media-enabled political information cycle described by Chadwick. At the core of this shift in mediated temporalities is a set of tactics designed to sustain networked and fleeting time-compressed communications across new and old media platforms—such as TV, the Web, handheld devices, and various social media platforms—and of course, mediated political dialogue, debate, and commentary (Gurevitch, Coleman &

Blumler, 2009). Methodologically speaking, the question to be answered is, why is there a need to study and analyze such dynamics in real time? One answer relates to the contingencies of interface time as a space that requires various strategies for communication (political communication in this instance) to be retweeted on the Twitter platform so as to recursively spread information across social networks and push the limits of socially mediated interface time.[3] In the context of political campaigns, public relations crisis management, or environmental disasters, such efforts to expand interface time take on an even greater significance as the emergent use of second screens, interfaces, and the interactive appendage to the broadcast sphere of political life (e.g., 24-hour news channels and live political programming) become increasingly important spaces to view immediate reactions to live events from a host of online political actors.[4] Such "live" or near–real-time reactions in the Twittersphere have consequently emerged as a site from which to support, ridicule, and/or refute the statements and claims made by public figures on live television. In other words, using political communication terms, micro-blogging sites such as Twitter have become key sites of "rapid-response"[5] to live political events and other particularly time-sensitive news stories.

Efforts to develop a live research paradigm in new media studies must take into consideration the speed of communications writ large. Publishing one's political opinion online (on a blog, for example) is no longer subject to editorial delay (except for comments, of course). User-generated content can be posted in real time at the click of a mouse. Does it not make sense then to build live-media time (or interface time, in the case of the Twitter ticker) into research methods to understand the effects of such media platforms and networks? Social media are structured to visualize only near–real-time contributions, and as such, their temporalities, flows, and interfaces set the context upon which political communications and campaigns are enabled, deployed, and represented through the introduction of real-time architectures (back-end code) and interfaces (e.g., feeds and tickers). In short, the fleeting nature of not only networked communication, but also the ever-changing software code, interfaces, and APIs that facilitate such micro-blogging activity, requires a temporal rethinking of what it means to conduct research on contemporary political communications and campaigning.

Networked or Web 2.0 communications and interactivity are overdetermined by conventions of the present; uploading, sharing, commenting, downloading, renaming, importing, embedding, or seeding are all enacted or published in the moment with little to no delay. The very language of networked life, political or otherwise, amplifies the immediate while clearly ex-distancing the technological, political, and economic underpinnings of such networks. It is this latter phenomenon that needs to be understood through the lens of the "live."

Twittering a Debate

In order to better understand the link between social media's compressed interface time and second-screen interactivity in their aggregated role as re-mediator of live political discourse, the example of live research discussed herein details our collaboration with the news division of the Canadian Broadcasting Corporation (CBC) during the 2008 federal election in Canada.[6] This study focuses on the development and execution of a near–real-time analysis of political tweets during CBC's live English-language national television broadcast of the federal leaders' debate[7]—a key moment in Canada's national election campaigns. Given Twitter's relative infancy in Canada at the time,[8] our live approach to the election-night study was designed to capture an early adopter moment in ICT-enabled political communications—one that sought to determine the influence of Net-savvy political operatives and also the degree to which the Twitter platform served as an interactive space for real-time reaction and commentary to a live broadcast event.

Given the minority status of the governing Conservative Party in the Westminster-style Canadian House of Commons, a series of potential election-inducing legislative showdowns had occurred during the year preceding the election. During this period, we developed a series of research methods and tools that tracked the growing importance and impact of the Canadian political blogosphere, and we published our findings. Having received substantial media coverage of our research during the Ontario provincial election in 2007,[9] our research lab was approached by producers in the news division of the CBC. For the project dubbed "Ormiston Online" (after its lead reporter, Susan Ormiston), the CBC brought together staff from all of their key news divisions (radio, new media, local, national, and 24-hour TV) in order to better disseminate the stories produced by the team for the CBC's myriad of news-focused programs and platforms.

We were approached by the CBC to assist in the development of a public Web portal, Internet campaigning research, and on-air interviews, and to provide advice related to developing news stories during the campaign. Unbeknownst to our research group at the time, the CBC had designed the project as a dry run for their subsequent multiplatform news realignment. Meanwhile, our methods of collecting data (e.g., blog posts and YouTube videos of the main party leaders) had already been established, tested, and refined prior to the collaboration. Three times per week, our team produced a ranking and short qualitative analysis of the most-linked-to blog posts from a sample of all the self-defined partisan political bloggers in Canada,[10] and a similar ranking of the week's most-viewed YouTube videos related to the federal party leaders during the campaign.[11] For our collaboration with the CBC, data was collected and analyzed three times per week and formatted for publication on the CBC's website (cbc.ca). As part of our contribution, we wrote one or two paragraphs in accessible language to provide context for the findings, which typically involved providing analysis for why certain posts or videos were receiving so much attention online.

Our research into the impact of blogging and YouTube videos on the election campaign process served as the backbone of our contribution to the CBC's coverage of the Internet-based aspects of the campaign. The first half of the official campaign period had witnessed a series of Internet-based scandals, missteps, and other campaign-related shenanigans that our project helped shed light on through our social media research and its subsequent dissemination through the CBC's website and broadcast platforms. Executives at the CBC, Canada's only national public broadcasting corporation, were reportedly pleased, and pushed for more content, analysis, research, and coverage of Internet-bound, campaign-related goings-on.

The most challenging live research component of the collaboration involved the use of Twitter during the nationally televised debate. Days before the party leaders were to meet for the televised debate, the producers of the Ormiston Online project met at the CBC's corporate offices to plan how we might cover the debate. Our discussions focused on aligning the broadcast and social media screens to highlight the real-time discussions and debates initiated on Twitter that we believed would be responding to the comments, barbs, guffaws, and poignant zingers served up by the party leaders.

Collecting the Tweets

Unlike our research on Canadian partisan blogs that restricted its sample to opt-in, self-described partisan members of one of Canada's party-branded blogrolls (see Chapter Two), the Twitter debate-night project was a decidedly open-ended affair that called into question the means by which we would filter or otherwise collect our debate-related micro-blogging posts. Recognizing the limits of Twitter's compressed interface time and its real-time use as a form of audience debate and dialogue, our project sought to analyze not only the content but also the context—that is, the exact time it was posted. Axel Bruns' (2010) initial research on the use of Twitter during the 2010 Australian televised leaders' debate was similarly designed to compare posts with those directed at a popular food television program, seeking to determine and compare the social media activity of contrasting social interests. In this context, Bruns' use of hashtags (#)—the most common form of creating new feeds or thread-like vertical posts of tweets on similar topics, individuals, events, or issues—to filter and collect relevant posts for his two simultaneous televised programs served as a helpful comparative sample to our data collection. Given the infancy of Twitter use in Canada at the time of our collaborative project, the user convention and practice of hash-tagging content had yet to fully emerge and mature. As a consequence, no one hashtag could capture a representative sample of posts during the Canadian televised debate. The use of specific hashtags has emerged over time after much conversation, debate, and adoption.

Unlike Bruns' study, our live research project sought to merge two sets of data to pose both qualitative and quantitative questions. We were not seeking to determine only the quantity of tweets during the debate broadcast, or their numbers in context to other live events.[12] Instead, our project sought to determine the interplay between the broadcasted comments by the leaders and individual and aggregate reactions on Twitter. After demonstrating that there was indeed a correlation between specific rhetorical flourishes, issues, or lively exchanges among the party leaders during the live televised debate and audience members' Twitter posts, we then sought to establish how such exchanges were deployed tactically to expand both Twitter's limited interface-time and the subsequent reach of the fleeting posts.

We decided to cast a wide net to collect micro-blogging posts related to the live-broadcast debate. Forty-eight hours before the debate, the project's staff—both academic and CBC-based—promoted the use of the #ormistondebate hashtag. Since findings from our project and of course the debate itself were being broadcast by the CBC, they were keen to cross-promote and otherwise brand their broader coverage. Overall, our research deployed a mixed hashtag, a Twitter account name, and a formal party leader name search-term "basket" to cull together as large a sample as possible.[13]

In addition to these meta-tag and formal-name search terms, the project also made important use of both the content of the tweets and the time stamp or log that accompanied each post. Such time stamps made it possible to cross-reference Twitter posts with the time-stamped transcripts of the leader's televised comments (made available to us by CBC news). While it took mere seconds to collect the tweets during the broadcast, our analysis was delayed by about 10 minutes as we waited for the delivery of the transcripts from the CBC via email.

Since our analysis was to be used for the televised post-debate on CBC news, the debate night proved to be incredibly hectic as we collected data to depict the minute-by-minute activity in the Twittersphere. The data was charted (see Figure 1) and broadcast later that evening on CBC News. While preparing the charts for broadcast, our research team also referred to the transcript of the debate to correlate jumps in Twitter posts to specific moments in the televised debate. Although we did not have enough time or space to include representative tweets on the charts to demonstrate these findings, we provided these tweets to our designated head reporter, who then used them on the live post-debate news broadcast to qualify the spikes in Twitter activity shown on our chart. Our research concluded that the most active moment over the first hour of the debate on Twitter occurred at 9:32 p.m., immediately after the left-of-center New Democratic Party (NDP) leader Jack Layton turned to Prime Minister Stephen Harper and let loose the first zinger of the debate: "Where is your platform? Under the sweater?"[14] The corresponding set of tweets clearly demonstrates a largely phatic or parrot-like use of the micro-blogging platform, meaning users either simply posted "wow"-like exaltations or reposted Layton's one liner, or both.

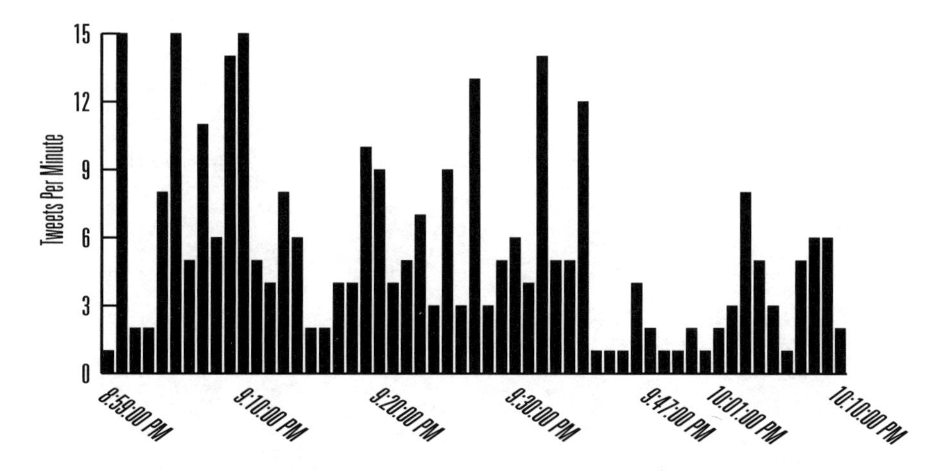

Figure 1. Twittering the Debate.

Reactions to the NDP leader's jibe also demonstrate a distinct partisan moment between political parties. The succession of twelve posts that repeated or otherwise exalted the witty one-liner over the next minute was only briefly interrupted by one tweet—from the centrist Liberal Party of Canada's campaign account:

> (9:32 p.m.) Liberal feed: Debates prove Jack Layton just doesn't get it.

Over the course of the evening, however, the Liberal Party was not the most active political party on Twitter. While many of the parties' well-known bloggers and online activists took turns supporting their respective leaders and taking apart the responses of their foes, only the New Democratic Party (NDP) actively prepared a rapid-response approach to Twitter on the debate night. Using the @JackLayton account, the NDP sent out a series of "fact check" tweets over the course of the 2-hour debate, posts that often included hyperlinks to more extensive rebuttals posted on the party's election website. The NDP, in short, used the medium to respond to their opponents' live statements in near–real time, adding a whole new temporality to the media spin that typically erupts at the conclusion of televised debates:

> (9:53 p.m.) jacklayton: FACT CHECK: Harper says he is making important investments in science and technology in Canada #ormistondebate

> (9:56 p.m.) jacklayton: FACT CHECK: Bloc not the only party with a Buy Canada policy—http://www.ndp.ca/page/7136.

While a number of users picked up on the tactic and lauded the party for its innovative use of Twitter, other comments suggested that viewers/Twitterers thought that Layton himself was posting the tweets while live on the TV debate set.

> (10:10 p.m.) @jacklayton, stop texting from under the table!

> (10:57 p.m.) @jacklayton explain to me how you are tweeting while the debate is on?

This confusion might be explained by the early stage of Twitter adoption and lack of established social media conventions, yet such strategic use of social media by a political party also highlights one manner in which media personalization is deployed during campaign events. Given that social media is built upon a lexicon and architecture of friend networks (as we previously noted in Chapter Three), the use of a personalized account by the NDP served to normalize partisan communications within the conventions of social media while at the same time extending Twitter's limited interface time onto their campaign website, where additional "fact checks," policies, and the party's campaign platform could be found.

A content analysis of the total number of mentions of the party leaders on Twitter, again performed on the night of the debate, concluded that while the NDP leader received substantial attention on Twitter (27 percent of all tweets mentioned Layton) during the course of the live broadcast, it was the first-time participation of the Green Party in Canadian debates that dominated the discussion on Twitter. Elizabeth May, the Green Party's leader, was mentioned in almost one-third (29 percent) of all the tweets collected during the debate night (see Figure 2).

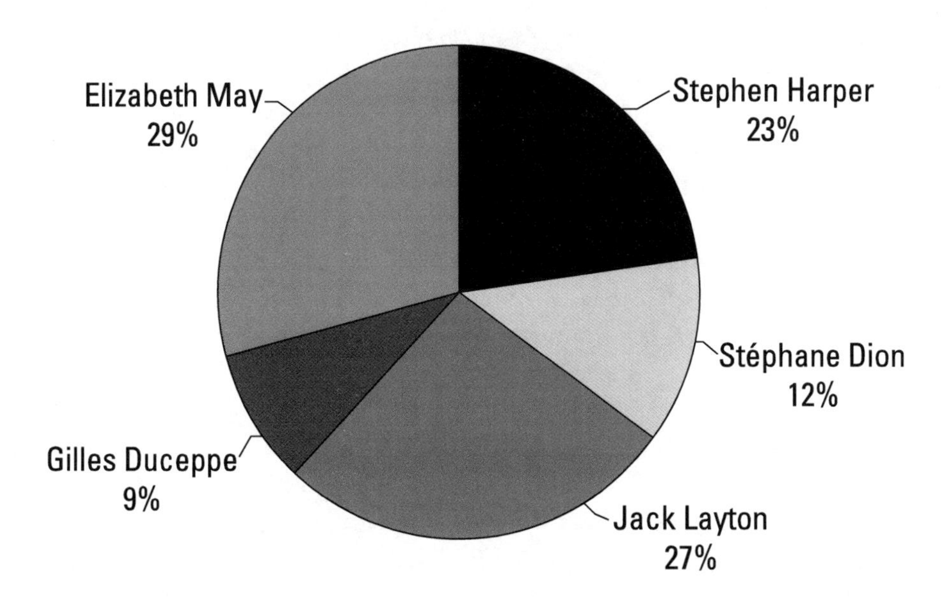

Figure 2. "Twittering about the Leaders." Broadcast October 2, 2008 on CBC national news.

Upon reviewing our data many months after the live research project, a series of other findings clearly stood out—evidence that supports and further qualifies the manner in which Twitter was used tactically by political parties, partisans, and other online viewers on the debate night. The multimediated nature of the debate evening, the interplay between viewership on the one hand and social media commentary and partisan campaigning on the other, is also amplified in a number of posts made during the debate evening. The Canadian federal leaders' debate happened to coincide with the live broadcast of the debate between the U.S. vice presidential candidates, which included the controversial yet media-friendly Republican nominee Sarah Palin. At the very outset of the Canadian debate, a number of users posted tweets referring to the use of multiple screens, online video streams, and the switching of TV sets to catch one or the other debate:

> (9:19 p.m.) Watching #vpdebate on CNN and #cdbdeb08 on CBC live stream #ormistondebate

> (9:21 p.m.) Just changed to the US VP Debate because so far it's better than watching Jack Layton and Elizabeth May attack @pmharper. Will go back soon.

Other users similarly engaged other social media platforms, in this instance the digital-photo hosting site Flickr, to capture and share their experiences of watching the Canadian leaders debate.

> (10:07 p.m.) The 5 leaders as they appear on my TV set. Elizabeth May http://flickr.com/photos/sarahroger/2908875540/.

Curiously absent, however, after retrospectively reviewing data from the debate night, is an expansion of Twitter's interface time onto other Web-based political documents. Apart from the previously noted NDP hyperlinks to their campaign website, of all the tweets posted during the two-hour live debate, only two tweets include links to other relevant Web-based platforms and objects. Such a finding seems counterintuitive given Twitter's predominant convention of sharing links to articles, YouTube videos, Wikipedia articles, and the like. While one of these posted links is rather whimsical, using a Web link to lyrics of a popular song to lampoon the NDP leader's choice of words,[15] the other is more tactically relevant in terms of expanding the sphere of the debate. At the outset of the debate, upon hearing the Green Party leader cite a report on the economy, a user finds the following document and shares it online:

> (9:18 p.m.) Here's a link to the OECD report Elizabeth May's talking about: http://www.oecd.org/dataoecd/33/55/40912642.pdf

Lastly, efforts to tactically manage—as opposed to perhaps simply expand—Twitter's interface time were also clearly evident in the hours leading up to the televised debate. A debate over an appropriate hashtag for the event quickly degenerated into partisan bickering and balkanization, with online Conservatives promoting the use of the hashtag #cdndeb08. There was, in short, a decidedly partisan and of course institutional element to various attempts at promoting specific hashtags, admittedly including the one used by the CBC's Ormiston Online project. Indeed, from the outset, some Conservative bloggers took offense at the CBC's promotion of the #ormistondebate hashtag, with some partisans questioning our respective role in this process:

> @greg_elmer... Did you play a part in setting up Ormiston to monitor the following twitter tag #ormistondebate?

Such after-the-fact findings, while further qualifying the expansion of both the time and space (screens, platforms) of micro-blogging during a live broadcast event, also highlight the limitations of real-time research, which cannot produce expansive, time-consuming reviews of data. Real-time or "live" research requires the pre-setting of a research agenda and a method of data collection, and, in this instance at least, a heavy reliance upon other forms of near–real-time comparative data (e.g., the debate transcripts provided by the CBC). Live research should therefore be viewed and understood as an effort at developing methods of collecting and analyzing data flows on platforms that hyper-accentuate the present, rather than simply enacting research and analysis in real-time.

Conclusions

The research-broadcast collaboration discussed in this chapter offered this book's authors an opportunity to intervene in public debates about the role that new media platforms play—or perhaps could play—in important social and political issues of our day; in this instance, in the very discourse enacted by Canada's political leaders. Scholars of new media suffer perhaps more than most in their frustrations at seeing their work—particularly time-sensitive research—delayed for many months and sometimes years. However, this is not a call to do away with established forms of peer review and scholarly publishing, but rather a proposal to question how new theories, methods, and venues for publishing and otherwise making research findings public can begin to address the growing importance of real-time media as a distinct event in itself (like a debate or media event such as a weather-related disaster), or a series of micro-events that in sum offer researchers insight into the structure and effect of "political cycles," as Chadwick notes (2011). Live research, as such, serves not only to question and understand the interface time of social media practices and platforms, but also to challenge the time-compressed and space-delimited sphere of academic scholarship.

Moving forward, such forms of live research need to distinguish themselves as research projects distinct from certain strands of information design—projects that seek to creatively visualize complex data sets and flows in the search for intuitive iconography and dynamic flux (Abrams & Hall, 2006). Live research, in other words, should not only be concerned with representing the world of things or their imprints, but also work to offer concepts, theories, and methods that might critically understand how users mobilize and sustain texts and other digital objects—by uploading, sharing, remixing, and downloading—across the field of networked communication. Live research could then serve as an important contingent step in recognizing the ever-shifting social media plane, and the subsequent tactics deployed to sustain meaningful communication in a socially networked media age.

In addition to recognizing live research as a new methodological approach to answer to some of the challenge of the acceleration of time on social media platforms, this chapter also argued for a new understanding of the temporal economy within which Web objects circulate. With the constant circulation of content promoted by platforms such as Twitter, the very notion of what constitutes political communication has to be revisited. Political communication is not about the rational exchange of ideas or the manipulation of the masses through rhetoric and propaganda. Rather, increasingly, political communication requires an understanding of tagged media objects—markers provided by users that indicate why particular objects and messages are worthy of sustained interest and visibility. More and more, the stamping—or as we call it in the next chapter, "tagging"—of objects is complicated by the superimposition of other platforms and media logics: Diverse black-boxed informational dynamics and constructions of the relationship between relevance and time have led to a new terrain of online political campaigning that will require further methodological inventions.

Notes

1. See Sysomos' September 2010 social media marketing study. Available at http://sysomos.com/insidetwitter/engagement.
2. See http://www.wired.com/epicenter/2010/10/its-not-just-you-71-percent-of-tweets-are-ignored/.
3. In this case we are talking specifically about Twitter; however, similar dynamics can be found on many other vertical-feed social media platforms.
4. By "political actors" we are referring to, for example, media pundits, political bloggers, politicians, and their staff.
5. For an early insider's view of the emergence of rapid response political tactics in the context of new information and communication technologies (ICTs), see Myers (1993).
6. The case study focused on Canada's fortieth general election. The campaign officially began on September 7, 2008 and ended on voting day, October 14, 2008. More details can be found on the Elections Canada website. See http://www.elections.ca/content.aspx?section=ele&document=index&dir=pas/40ge&lang=e.
7. Canadian convention for televised debates is typically to broadcast in both of Canada's official languages, English and French. This study focuses exclusively on the English-language debate broadcast on October 2, 2008, though a dry run of our methods were informally tested during the French-language debate held the day earlier.
8. The platform launched worldwide in July 2006.
9. See http://www.infoscapelab.ca/ontarioelection2007.
10. An archive of the Ormiston Online project can be found at http://www.cbc.ca/news/canadavotes/campaign2/ormiston/.
11. At the time, YouTube provided only total cumulative views of videos. Working with the platform's API, we wrote a software script that determined on a weekly basis how many views a video received.
12. Though as we shall see, a number of the posted tweets made reference to switching back and forth between the Canadian and U.S. vice presidential televised debates.
13. The search terms and hashtags included: #ormistondebate, the Twitter account names for the party leaders and campaigns (i.e., jacklayton, LiberalTour, Pmharper, ElizabethMay, gillesduceppe), and the formal names of the federal party leaders ("jack layton," "elizabeth may," "gilles duceppe," "stephane dion," and "stephen harper"). The total sample included 558 tweets.
14. The comment made reference to the Conservatives' lack of a formal party platform, and a party advertisement depicting the Conservative prime minister in an atypically informal sweater.
15. "I'm sure it's a coincidence but Jack Layton just paraphrased a Propagandhi song."

References

Abrams, J. & Hall, P. (2006). *Else/Where: Mapping New Cartographies of Networks and Territories.* Minneapolis: University of Minnesota Design Institute.

Allan, J. (ed.) (2002). *Topic Detection and Tracking: Event-based Information Organization.* Norwell, MA: Kluwer Academic Publishers.

Anstead, N. & O'Loughlin, B. (2011). "The Emerging Viewertariat and *BBC Question Time*: Television Debate and Real-Time Commenting Online." *The International Journal of Press/Politics* 16 (4), 440–462.

Bruns, A. (2010). "Politics vs. Masterchef: The View from Twitter." Available online at http://www.mappingonlinepublics.net/2010/07/26/politics-vs-masterchef-the-view-from-twitter/.

Chadwick, A. (2011). "Britain's First Live Televised Party Leaders' Debate: From the News Cycle to the Political Information Cycle." *Parliamentary Affairs* 64 (1), 24–44.

Chadwick, A. & Howard, P. (2009). *Handbook of Internet Politics.* London: Routledge.

Cunningham, S. D. (2008). "Political and Media Leadership in the Age of YouTube." In P. Hart & J. Uhr (eds.), *Public Leadership: Perspectives and Practices* (pp. 177–186). Canberra: ANU E Press.

Elmer, G., Langlois, G., Devereaux, Z., McKelvey, F., Ryan, P. M., Redden, J. & Curlew, B. (2009). "Blogs I Read: Partisanship and Party Loyalty in the Canadian Blogosphere." *Journal of Information Technology & Politics* 6 (2), 156–165.

Elmer, G., Ryan, P. M., Devereaux, Z., Langlois, G., Redden, J. & McKelvey, F. (2007). "Election Bloggers: Methods for Determining Political Influence. *First Monday* 12 (4). Available online at http://firstmonday.org/htbin/cgiwrap/bin/ojs/index.php/fm/article/view/1766/1646.

Fuller, M. (2003). *Behind the Blip: Essays on the Culture of Software.* New York: Autonomedia.

Gurevich, M., Coleman, S. & Blumler, J. G. (2009). "Political Communication—Old and New Media Relationships." *The ANNALS of the American Academy of Political and Social Science* 625 (1), 164–181.

Galloway, A. (2004). *Protocol: How Control Exists after Decentralization.* Cambridge, MA: MIT Press.

Geere, D. (2010). "It's Not Just You: 71 Percent of Tweets Are Ignored." Available at http://www.wired.com/epicenter/2010/10/its-not-just-you-71-percent-of-tweets-are-ignored/.

Harman, G. (2009). *Prince of Networks: Bruno Latour and Metaphysics.* Melbourne: Re.Press.

Hine, C. (2007). "Connective Ethnography for the Exploration of e-Science." *Journal of Computer-Mediated Communication* 12 (2), article 14. Available at http://jcmc.indiana.edu/vol12/issue2/hine.html.

Jungherr, A. & Pascal, J. (2011). "One Tweet at a Time: Mapping Political Campaigns through Social Media Data." Paper presented at the 6th ECPR General Conference, Reykjavik, Iceland.

Kluver, R., Jankowski, N., Foot, K. & Schneider, S. (eds.) (2007). *The Internet and National Elections: A Comparative Study of Web Campaigning.* London: Routledge.

Lash, S. (1999). "Objects that Judge: Latour's Parliament of Things." *Transversal.* Available at http://eipcp.net/transversal/0107/lash/en.

Latour, B. & Weibel, P. (2005). "From Realpolitik to Dingpolitik—or How to Make Things Public." In B. Latour & P. Weibel (eds.), *Making Things Public: Atmospheres of Democracy* (pp. 4–31). Cambridge, MA: MIT Press.

Markham, A. N. (1998). *Life Online: Researching Real Experience in Virtual Space.* Lanham, MD and Oxford, UK: Rowman & Littlefield.

Miller, V. (2008). "New Media, Networking and Phatic Culture." *Convergence: The International Journal of Research into New Media Technologies* 14 (4), 387–400.

Myers, D. (1993). "New Technology and the 1992 Clinton Presidential Campaign." *American Behavioral Scientist* 37, 181–184.

Norris, P. (2000). *A Virtuous Circle: Political Communications in Postindustrial Societies.* Cambridge, UK: Cambridge University Press.

Rogers, R. (2006). *Information Politics on the Web.* Cambridge, MA: MIT Press.

Schneider, S. & Foot, K. (2010). "Object-Oriented Web Historiography." In N. Brugger (ed.), *Web History* (pp. 61–82). New York: Peter Lang.

Thelwall, M., & Stuart, D. (2007). "RUOK? Blogging Communication Technologies During Crises." *Journal of Computer-Mediated Communication* 12 (2), article 9. Available at http://jcmc.indiana.edu/vol12/issue2/thelwall.html.

CHAPTER 6

Networked Campaigns: Traffic Tags and Cross-Platform Analysis on the Web

Politics has always been about networking. Before seeking office, prospective candidates are advised to identify well-connected individuals—those who can help raise funds, make insider connections in party circles, and otherwise "open doors." While political networking today still requires face-to-face meetings, it now also requires a virtual dimension—one that raises significant opportunities and pitfalls for campaigns and political life in general. For candidates, political party strategists, and communications staff, social media platforms (such as the previously interrogated Facebook, YouTube, and Twitter) offer distinct opportunities to reach segmented communities and to narrowcast messages to party members and partisans in specific electoral ridings and districts, regionally or nationally, at particular times of the day. Social media can also be used as a promotional tool for specific purposes (campaign stops, stump speeches, fund-raising, candidate nomination meetings, leadership contests, etc.). At the same time, however, social networking sites challenge the ability to control and otherwise manage so-called talking points,[1] election policies and platforms, and overarching election campaign "scripts." A message, image, or video can be shared with political opponents and remixed or critiqued in very short order, or in near-real time (as discussed in Chapter Five).

Networked political communication, in other words, has become mutable, evasive, and much more difficult to manage in the social media universe. For political actors, the sheer number of new media spaces—Web 2.0 platforms, social networks, and information aggregators—complicates the ability to deploy contemporary political campaigns. Where does one start? What should one share or reserve solely for party supporters? How should one respond to political attacks and rumors on social media? Gone are the days when political strategists focused exclusively on editorial boards of newspapers, briefing notes, fund-raising letters, and stump speeches for media campaigns' buses and planes.

In light of such radical changes in the information and communications sphere, political scientists and communication scholars have sought to develop new experimental methods of understanding the digitally networked terrain of politics (Chadwick & Howard, 2008). Web 1.0 studies—primarily concerned with political communication, organizing, and networking on the World Wide Web—sought to develop methods of mapping hyperlinked relationships among websites for political candidates, parties, and civil society organizations, to name a few. Building upon earlier forms of social network analysis, hyperlink networking methods and tools (Park & Thelwall, 2003)—notably, programmable "crawlers" that jump from link to link—were used to identify key political actors or "hubs" in networked hyperlink diagrams (Rogers, 2002; Garrido & Halavais, 2003; Hindman, 2009). This research involved locating the most influential political actors on the Web by identifying the most-linked-to Web pages. For our purposes, we refer to such forms of analysis as "http methods," in recognition of their use of a single form of code that links together html documents on the World Wide Web: the HREF (or hyperlink) command (Foot & Schneider, 2006).

The reliance upon the hyperlink as the sole indicator of techno-political associations both online and offline (Foot & Schneider, 2006, p. 38; Rogers & Marres, 2000), however, has not been without its skeptics (Jankowski & Selm, 2008). Elmer (2004) has argued that hyperlink mapping faces numerous technological hurdles as Web servers often crash and need to go offline for routine maintenance. Websites and pages are often blocked for a host of other reasons political, interregional, or proprietorial (see also Chapter Three's discussion of Facebook). Researching political connections and associations on the Web thus requires one to recognize disconnected or disrupted forms of networked computing. Whatever one thinks of early forms of hyperlink analysis, these methods clearly contributed to innovative forms of data visualization, as attempts to more accurately—or perhaps more creatively—represent distributed forms of political networking.[2] Yet, new data visualization software,[3] some representing a seemingly infinite number of hyperlinks,[4] also produce undecipherable, Death Star–like maps of hubs and spokes, posing significant challenges to meaningful forms of analysis (Bourret et al., 2007). Moreover, hyperlink maps render and visualize only *functional* hyperlinks and websites at specific moments. Where are the network maps, for example, that denote disconnections, server timeouts and crashes, and deleted links between sites? Such positivism—meaning successful and empirically verifiable links—in the absence of various forms of disconnectivity and dysfunction serves to reify political networking as *successful forms of connectivity*. Political networking (much like computer networking) is quite often the opposite: labored, unstable, precarious, unverifiable, sometimes unconscious, and hidden. How might emergent forms of research therefore acknowledge such qualitative distinctions in and across networks?

This chapter sets the stage for an approach to the study of Internet politics and networking that addresses the impact that new Web 2.0 interactive platforms have upon what we refer to as "the conditions of networked connectivity." By *conditions*, we are suggesting that connectivity itself has been largely understudied, or worse—interpreted as either a sign of political alliance, support, or merely "successful" connection. As such, we offer building blocks for methods that attempt to account for connection failures, disruptions, and roadblocks; some that are "accidental," and others that are the obvious result of restrictive terms of use encoded into Web 2.0 platforms (and their application programming interface, protocols, and algorithms). By focusing on the conditions of connectivity, we seek to integrate user-based experiences and, of course, their shared, remixed, and uploaded digital objects[5] into the broader research paradigm.[6] This research agenda involves mapping networking opportunities and restrictions as well as potential dysfunctions and incompatibilities.

In developing new methods for studying the relationship between political actors, objects, and platforms online, this chapter presents a brief "meta-tag" analysis of political keywords on the World Wide Web as a test case for demonstrating how non-hyperlinked forms of software code can also provide insight into networked political campaigns on the World Wide Web. After some initial reflections and analysis of our "tag"-based study of political networking, we discuss how such tags operate in the much more complex world of Web 2.0, where users are increasingly called upon to self-categorize (through titles, keywords, hashtags, etc.) their online contributions (e.g., images, blog posts, tweets, comments, videos, etc.). We conclude with an effort at expanding and analyzing how a plethora of Web 2.0–based forms of user- and automatically generated software code can be harnessed to better understand the possibilities and constraints of political networking—cultivating a cohesive political campaign—across a number of websites and Web 2.0 platforms. The ultimate goal of this longer-term project is to offer methods and tools that might diagnose the possible reach of online political campaigns, communications, and networks. Our approach seeks to determine the constitution and constraints afforded by different sets of relationships among uploaded and shared Web objects, political actors, and Web-based platforms.

In moving from so-called Web 1.0 "http" or "html" approaches to Web 2.0 cross-platform–based methods, we are particularly interested in harnessing—methodologically speaking—user-generated forms of classification, or *tags* to use the Net-vernacular. These tags, or "keywords," are commonly used by social media partisans and activists to associate their online contributions with like-minded political and social debates, actors, sites, platforms, and other online objects. To identify the relationships forged by objects, actors, and platforms, this chapter also makes the case for identifying discrete forms of communication and networking *in motion*; that is, as Internet-based network *traffic*. While previous forms of http-based hyperlink analysis offered a means of identifying relationships among websites and their assumed owners/Webmasters, our *traffic tag* approach seeks to determine the multiplicity of avenues across Web 2.0 platforms—or conversely, the dead ends—that limit the reach and political possibilities of online campaigns. Only through tracking the unique forms of ID associated with platforms (e.g., through their URLs), online political actors (e.g., their accounts, usernames, etc.), or networked objects (e.g., titles, URLs, etc.) can we begin to diagnose the possibilities and pitfalls of Web 2.0–based political networking, communications, and campaigning.

Trafficking Political Rhetoric: "Stand Up for Canada"

In this section, we offer a brief analysis of how meta-tag keywords on the World Wide Web can be harnessed and analyzed to understand the reach and circulation of political campaigns on the Internet. The study offers a glimpse into why subsequent 2.0 forms of analysis need to take into consideration the role that self- and automatically generated tags play in the generation of possible avenues for networked political content and actors across a number of popular Web 2.0 platforms. So as not to overstate the novelty of our proposed method of research for mapping political networks, issues, actors, and objects across the 2.0 universe, it is important to note that the building blocks of a more nuanced Web 2.0–enabled form of network mapping or "traffic tags" approach to the study of political campaigns were to a much lesser degree already present on the World Wide Web. While html-encoded Web pages offered href tags (hyperlinks) as conduits for network mapping, http-header meta-tag keywords and other meta-data have afforded other opportunities for qualifying and expanding network analysis.[7] One such line of inquiry has focused on the relationship between websites and their visibility and ranking via industry-leading search engines. Google's indexing bots, for example, "read" the http-header keywords of html Web pages so that they can be better integrated into Google's archival, page-ranking, information aggregation, commercial advertising, and user-profiling functions (Halavais, 2008). Webmasters thus encode their websites' header keywords to sufficiently represent their sites' content, thereby enabling accurate indexing from Google and other information aggregators. Such keywords link websites to Web aggregators, most notably Google via its "page-rank" algorithm (Brin & Page, 1998).

Political consultants and campaign staff in the 2004 U.S. presidential election were quick to recognize the many different techniques that campaigns could use to better "optimize" their candidate's visibility on the Web by refining titles and other keywords in the headers of campaign Web pages (Easter, 2008). Similarly, the homepage for the Conservative Party of Canada includes rather obvious meta-tag keywords such as "Conservative Party" and "Stephen Harper" (the Canadian prime minister). However, reviewing the http header—which one can easily do by choosing the "view > page source" pull-down menu on most Web browsers—also reveals the strategic insertion of a recent election campaign slogan, "Stand up for Canada," and a short list of political issues and buzzwords: "trades, transit, accountability, childcare, etc."[8] While Conservatives in Canada strategically use such tags to brand their political campaigns and messages, Webmasters can dream up and encode their http header with any sequence of keywords. These keywords are tactically deployed to gain greater visibility on Google, as a higher ranking ultimately results in increased traffic to their site.[9] In lieu of considering these connections between websites as networked associations, we should consider that such keywords serve to self-identify Web pages and cultivate new sources of traffic. The "tagging" of one's content—through the use of keywords—suggests a degree of self-promotion; a form of publicity that from time to time stretches the indexical purpose of such meta-tag keywords.[10]

A brief search for the "Stand Up for Canada" phrase, using the Google search engine, offers a glimpse into the circulation and adoption of such politically loaded and "genetically" encoded[11] words from the Conservative Party's website. Google results for the Conservative's "Stand Up for Canada" provides an intriguing picture of the numerous websites and Web 2.0 platforms that repeat, adopt, or otherwise circulate the phrase.[12] In addition to a page from the Conservative Party website that reuses the phrase as a generic headline for political reaction to the constitutional crisis that emerged shortly after the Canadian federal election in 2008 (see Chapter Two), Google also returns the following results:

Web Platform	Content
1. Conservative Party Website	Political content using phrase as headline
2. Conservative Party Website	Home page; phrase used as main header-banner
3. Childcare Resource Center	Archive of Conservative Party platform document that used the phrase in its title
4. YouTube	Two YouTube videos: (1) one critical of the PM, using phrase in title and in content; and (2) phrase included in title and description of a video critical of North American union policies
5. United Steelworkers Website	"Stand up for Canada: Save Manufacturing" advocacy article
6. Political Website	Uses the phrase to critique a wide set of government policies
7. Personal Blog	"Time for CRTC to Stand up for Canada" title for blog post
8. Prime Minister's Facebook Page	Headline to same article as result #1, reproduced for Facebook
9. Government Web Page	Speaking notes for government minister that uses phrase in title and three times in body of speech
10. Prime Minister's Myspace Page	Reproduction of results for #1and #8

Figure 1. Google Returns for the Phrase "Stand Up for Canada," Organized by Platform and Location of the Phrase.

With this brief glimpse of meta-tag keywords one can make a series of preliminary though important methodological conclusions and claims. The broadest of all supports our contention that certain Web-based tags—words inserted into a http header by Webmasters in this 1.0 case study—can be used in much the same way that hyperlink analysis has been deployed; that is, to track the relationship between and dissemination of digital objects, coordinated campaigns, and lastly, political actors. While the nature of digital objects tends to multiply exponentially in a Web 2.0 environment, a keyword- and tag-based method of analysis—such as the one conducted above—is largely restricted to the study of plain text, political keywords, or short catchphrases used to symbolize ideologies, policies, and legislative priorities. However, by tracking, albeit rather simplistically, the dissemination of such keywords across the Web—as aggregated by Google—we can also catch a glimpse of the spread and adoption of such political keywords and slogans (e.g., "Stand Up for Canada"), the actors involved, and the political contexts and platforms used. For example, five of the top ten results of the phrase aggregated and ranked by Google emanated from either the government of Canada or its ruling political party (the Conservative Party). Results #1, #8, and #10 (see Figure 1) clearly demonstrate a coordinated cross-platform campaign by the Conservative Party to utilize a title ("Stand Up for Canada") to frame a verbatim attack on their political opponents. Result #2 suggests that the Conservative Party is also using the heading as a more generic keyword to frame its broader public relations strategy. The ninth result, where the phrase is found in the text of a speech delivered by a Conservative government minister, demonstrates that the phrase "Stand Up for Canada" is used not only for partisan purposes but also as a key political phrase repeated in public and policy settings. The third result for the phrase points to the phenomenon of third parties, in this case a childcare resource center, archiving certain governmental and political documents for, presumably, their own political use (such as lobbying purposes and internal membership campaigns). Opponents and critics of the Conservative government are equally accounted for in Google's top ten results for the meta-tag phrase "Stand Up for Canada." Two user-generated 2.0 sites—a blog and a YouTube account—clearly attempt to usurp the government's phrase for critical purposes, as does, to a lesser extent, a manufacturing advocacy piece from the website of the United Steelworkers trade union (USW).

From this brief analysis of embedded html keywords, then, one can clearly see that the political phrase "Stand Up for Canada" is a contested one online, bringing together party communications staff, government departments' ministers, interest groups covering industrial and social issues (e.g., steelworkers and childcare advocates), and social media users. This brief analysis shows that the phrase circulates across established html websites to blogs, top English-language social networking sites such as Facebook and Myspace, and the popular YouTube social media aggregator. Objects, actors, and political campaigns become increasingly re-mediated across social media and Web 2.0 platforms, and as such, the need to develop a traffic-tags approach to the study of political networking takes on an even greater sense of urgency.

Social Media: The Sharing of Objects

Since much of this chapter presumes a radical shift in Web operability (from 1.0 to 2.0), some important conceptual remarks on social media are required to expand on this previous brief test-study to establish the building blocks of a 2.0 method for researching political networking. This is particularly urgent because, as a concept, Web 2.0 feels a bit like a black hole: everything gets trapped within its porous boundaries, from commercial and private social networks to the collaborative site Wikipedia; from the latest online social networking craze, Twitter, to one of the first and most enduring successful online business models, Amazon.com (O'Reilly, 2005).

Mainstream discourses about Web 2.0 often refer to a projected perception of the state of the contemporary World Wide Web as correcting the shortcomings of the previous Web 1.0 era and fostering a democratically infused and dis-intermediated commercial sector (Eysenbach, 2007; Bakos, 1998). Thus, while YouTube, Facebook, and Wikipedia each emphasize different functions, media, and business models, all rely heavily upon user-generated content. To clarify, Web 2.0 platforms rely on their users not only to produce and upload content, but also, more importantly, to share and circulate it across friend networks of like-minded individuals and groups. Social networks are, in effect, produced by the sharing of objects on their sites. Without the trafficking of objects (e.g., links, images, videos, text, etc.), the owners of such sites would be unable to aggregate and data-mine personal information from users and their like-minded friends. Similarly, the popularity of YouTube is not simply linked to its capacity to act as a repository or archive of videos; rather, it continues to grow as a result of its ability to share, through embedded code, videos on a number of platforms across the Web (Green & Burgess, 2009).

Web 2.0 relies upon these shared objects as well as avenues for circulating said objects that link together individual users and their networking affinities. We like to think of these objects and avenues as "friendly traffic"; of course, this is not to downplay the fact that such sites subsequently aggregate users' psychographics, profiles, and online behaviors to sell "targeted" advertising (Warschauer & Grimes, 2008). The focus on friend-based traffic—the sharing of objects on and across social media platforms—thus calls into question the architecture of social media sites as objects of research and analysis. Political research on social media must take into account not only users (be they political partisans or institutions) but also the possibilities that social media platforms afford on their sites—the opportunities and roadblocks of uploading, sharing, and networking across the Web, handheld devices, and beyond.

2.0 Networking: From Universal Protocols to Unique Identifiers

To begin to map and analyze the circulation of objects, actors, and broader networked campaigns on the Web today, we argue for a cross-platform approach—a method that seeks to determine the networking opportunities and limitations among and across Web 2.0 sites. A method that would witness the unfolding of the circulation of virtual political campaigns and networks via Web 2.0 platforms would be of considerable benefit to researchers in terms of identifying specific networking opportunities, limitations, and pitfalls in the political sphere. The first step in developing such a perspective requires moving beyond and below the user interface. That is, we need to challenge our perception of the Web as rooted within the visual aesthetics of the user interface. This is all the more crucial and challenging on proprietary and closed websites such as Facebook, where the interface becomes a limiting factor because our only point of entry is through the customized or "personalized" (first-person) perspective of our own networked environment. Web 2.0 social networking is an intensely personalized medium; as we noted in Chapter Three, no two Facebook interfaces and accompanying "friend" networks are the same. We all see, and operate within, Facebook through the contours of our own social networks. Such networks bias—and to a degree, determine—the actions we perform via Facebook's search functions, skewing the results to highlight our own aggregated friend-network profiles.

In short, no two search results via Facebook are alike—even for the exact same search term. Thus we can never have access to the totality or even a common set of information available on Facebook via the user interface—and as network researchers, this always-already personalized interface and algorithm complicates our ability to analyze from third-person perspectives; that is, from the "outside." Indeed, the user perspective creates an oddly narcissistic worldview of Web 2.0—one individuated through a me-centric (and thus uncannily familiar) network interface. Adopting a cross-platform perspective helps to overcome the limitation of the user's worldview by disaggregating objects, actors, and networks from 2.0 user accounts.

Web 2.0 protocols are largely concerned with managing users and user-generated content, or "objects"; these are the connections that enable relationships that populate networks across Web 2.0. Web 2.0 platforms set up the channels through which information can circulate. Our proposed method seeks to develop tools to track, map, and visualize such channels or traffic routes. This approach has roots in the critical aesthetics of software studies. For instance, Matthew Fuller's *Webstalker* (2003) offered an alternative Web browser designed to represent the linked relationships between websites; it was a browser devoid of any aggregated information or iconic graphics. Our critical approach to Web 2.0 platforms likewise requires a process of disaggregating the relationship between interfaces and back-end code and protocols—a form of reverse engineering, if you will. The building blocks of a disaggregated Net, as previously stated, begin with a process of identifying the key components in political-computer networking—actors, objects, and platforms—each of which contains unique forms of ID, including user-generated tags. Once we can identify each of these actors and objects on the Net, we can then map the traffic, or the routes, of such ID tags to determine how and where political campaigns circulate across the Web.

Traffic tags serve not only to organize cross-platform communication but also to enable connections across different actors and to organize online activity. Our focus on traffic tags emerged from a realization that there is a need to include the "beyond and below" discursive dimension of online content, and from an acknowledgment that what used to be discrete Web objects have morphed into entities capable of enabling different forms of simultaneous connection at different levels. With regards to the beyond-and-below discursive dimension of online content, we mean the material aspects and social effects of political content networked across Web 2.0 platforms. "Below" encompasses the data processes and network routes through which online content is circulated and published. "Beyond" refers to the capacity of content to represent, but also, and more crucially in the online political context, to organize and spur action (e.g., voting, fund-raising, protesting). Furthermore, the acknowledgment of the morphing of Web objects into traffic tags offers a methodological incentive to pay closer attention to the beyond and below aspects of online content.

By way of example, Barack Obama's famous political phrase "Yes We Can"—as a slogan, a rallying cry, a lasting rhetorical gesture and the summation of an expansive and expensive political campaign—should be considered as a brand; that is, as a "platform for the patterning of activity, a mode of organizing activity in time and space" (Lury, 2004, p. 1). There are several critical aspects of patterning and organization expressed through the online circulation of "Yes We Can." First, "Yes We Can" is a multidimensional Web object: it is a rhetorical logos, a cultural symbol to which a range of media objects (official texts, videos and pictures, citizen responses, critiques, and parodies) are associated. As a link object, it is also a deictic sign, or pointer (Shields, 2000; Elmer, 2006), to different platforms (e.g., the official campaign website, the Facebook page, and other websites). Underlying its repurposable form as a button that can be embedded in individual Facebook pages, blogs, and websites, it represents political action as a declaration of allegiance and voter intention. As an application—especially, as a Facebook application developed by Obama's campaign staff—it serves as a covert polling technology to cull more information on supporters and would-be voters. As such, "Yes We Can" is a multilevel traffic tag that serves to organize and centralize different types of activity.

From the point of view of the user, "Yes We Can" is both content and a deictic pointer to a broader community of like-minded individuals. At the political level, the importance of the "Yes We Can" logo is not simply that it is a symbolic rallying cry, but that it is also an operative one that can quantify its effects by being turned into a tracking device that enables precise quantification of the reach of a message. From a computer-networking point of view, "Yes We Can" is the user-understandable facet of a range of data processing that aims to identify and link relevant information according to different platform logics. For instance, while the Google search engine logic used to identify the most relevant material for the large population of users, the Facebook search engine operates through the logic of personalization, by gauging friends' preferences and geographic proximity. Traffic tags are thus operators that allow for the connection of different actors via, for instance, political rallying and Web tracking. They express multiple practices that aim to organize political relationships, political discourse, and informational networks. For this reason, traffic tags should be considered as objects of analysis to better understand political activity across Web platforms, as well as analytical objects through which we can derive new methods for tracking the unfolding of online political campaigns, communications, and networks.

Traffic tags can be human-generated, such as the title of video or the formal name of a user as they appear on the user interface, or the user tags that describe how an object belongs to a particular class of object (e.g.,"X's wedding" or "election 2008"). Traffic tags can also be computer-, software-, or platform-generated; for example, the unique identification numbers assigned to YouTube videos and to users on Facebook. Traffic tags allow for the identification of objects across the Web, most notably through search engines, but also through application programming interfaces (APIs), which govern how objects circulate within and sometimes across most Web platforms. For instance, when a user clicks on the "Share on Facebook" button after watching a video on YouTube, the ID number of the video will reappear in the Facebook source code of the user's page. The current challenge thus lies in identifying and following traffic tags associated with Web objects in order to see how information circulates within and across Web 2.0 platforms. The process of tracking the migration of object- or actor-specific code will provide us with clues about how cultural processes that are traditionally visible only at the level of the user-interface are governed by the largely commercial imperatives of APIs, particularly on the larger and more popular platforms such as Facebook and YouTube.

The Taxonomy of Traffic Tags

While meta-tags offer an important contrasting view to the use of hyperlinks as indicators of political associations and networks, their use has been vastly complicated and expanded in the Web 2.0 universe. In fact, as we have argued elsewhere (Langlois et al., 2009), such forms of user-generated content serve as a key component in the production of Web 2.0 sites since they rely almost entirely upon user-generated content to function and thrive. However, the task of developing methods for tracking individual users and networked political objects across platforms is a complex one, in large part because each platform has its own set of protocols that disrupts the more free-flowing aspects of Web 1.0 (or html-based forms of publishing and networking). In the remainder of this chapter, we identify new forms of code and software functions that might allow for the tracking of objects and users across Web 2.0 sites. Such software artifacts serve as possible sites of 2.0 research, though this decentered method of analysis, which begins with objects and users as opposed to networks, communities, or other digital collectivities, will inevitably raise questions about one's choice of a starting point.

What is the rationale for choosing the objects one begins to track, and what sequence and series of information aggregators does one deploy to view the dissemination of traffic tags? Before we move on to discussing such new trajectories of research, we should reiterate that traffic tags typically come in two forms—both of which are required to track objects and map routes of networked content, and relationships between users, content, and other users. Namely, there is code that individually identifies specific users/objects, and code that facilitates the circulation of shared objects. This method is not entirely new, as it also duplicates—albeit with some differences—the techniques and technologies that are deployed to diagnose the circulation of commodities, consumers, and services in today's economy (Elmer, 2004). In lieu of traffic tags, then, such networked objects, users, and routes have employed well-known technologies such as barcodes, RFID tags, and "just-in-time" delivery systems for many decades now.

We have identified a number of traffic tags that exemplify our search for code that can be employed in an object-centered method of Web 2.0 analysis.

This list is by no means exhaustive; rather, it is meant to offer a starting point for discussion:

- plain language (text)
- individual user IDs
- APIs
- tags that accompany user-generated objects (self-generated, auto-generated)
- hyperlinks
- spam strings
- RSS feeds
- object titles
- file formats
- usernames
- formal names
- IP addresses
- copyright code (e.g., creative commons)
- email addresses

Plain language, or text, is one the most overlooked forms of traffic tags on the Web. As we argue elsewhere with respect to the reuse and circulation of Wikipedia entries (Langlois & Elmer, 2009), one can take formal language and deploy it in a series of Net-based information aggregators (search engines, for example) to identify the dissemination of similar or exact duplicates of sentences and paragraphs. Plain language is a particularly cogent form of traffic tag because it doubles as both a semiotic and deitic sign (Elmer, 1997; Shields, 2000), meaning that it can provide researchers with the rhetorics of networked politics as well as insight into how terms are used to take users—literally, in the case of hyperlinked words—to other documents and Web platforms.

Application program interfaces (APIs) are similarly pivotal in our proposed research since they sit "in-between" interfaces and back-end code, often providing more savvy users with the ability to data-mine specific platforms for information on users and objects. APIs serve as search engines of sorts, as they link together users with objects and particular spaces on platforms such as Facebook (e.g., on groups or "causes" pages, etc.). One can "query" an API, for example, for various data associated with a particular user[13] or group of users. APIs can also be used to better understand how networked political objects move across platforms, and are slightly modified or become the domain of specific 2.0 platforms, to the degree that their sharing becomes more difficult.

Really Simple Syndication (RSS) feeds similarly offer researchers a universally recognizable code embedded on many political websites, blogs, and media sites. They serve in many respects as a content portal—a mega-hyperlink in 1.0 language—to the extent that they create a gateway from which almost all content, and indeed some meta-tags and information for specific website entries, stories, or posts (e.g., date stamps, bylines, etc.), can be collected and used for comparative cross-platform analysis. Much like APIs, in other words, RSS feeds serve to traffic meta-tagged content. Our own analyses of political blogs in Canada used the RSS feeds from partisan blogs to perform various forms of content analysis across the Canadian political blogosphere (Elmer et al., 2009). A slightly modified version of these traffic-focused tags and code includes the creative commons logos and tags, as signs and code that govern, classify, and enable access to various forms of multimedia on the Web (Flickr images, for example). Content, actors, and platforms associated with creative commons licenses speak directly to the rules concerning the ability to publicly use, reuse, remix, and broadly share digital objects. Locating creative commons code across platforms using search engines, APIs, and RSS feeds thus provides helpful sets of data for gaining insight into the various forms of digital ownership and subsequent trafficking of content that take place across the Internet. Such issues are of increasing importance for the political sphere as various jurisdictions around the world move to more open-source models of information management and access.[14]

Identifying and tracking the contributions of political actors (e.g., partisan bloggers, vloggers, political staff, journalist-bloggers, etc.) is perhaps one of the easiest components of our suggested method of inquiry. Because almost all Web 2.0 sites require some form of user registration, individual IDs are commonplace. These IDs are platform-specific, which can help when trying to determine the success or failure of various cross-platform political campaigns. User accounts almost always require the registration of a unique username, thus making it relatively easy to track the content and objects uploaded, remixed, and commented upon by specific users. This logic might also be extended to include less formalized definitions of usernames: for example, "aXXo," a well-known user of peer-to-peer software known to upload and circulate DVD-ripped material on bit-torrent networks.[15] Likewise, email addresses offer opportunities to identify the circulation and contribution of individuals across platforms and timeframes, though with important caveats that speak to the limits of political networking both as a practice and a site of research. Emails listed on Facebook pages are not retrievable through interface searches or through the platform's API, thus making it harder to analyze and circulate calls to action posted on Facebook that often end in an organizer's email address. IP addresses are one of the more reliable and unique forms of identifying specific users, or in the case of "whois" searches, the unique address where a computer is registered. Journalists often turn to "whois" searches during election campaigns to determine the owners of specific attack or parody websites—a daunting task, as according to one estimate, more than 2,357 sites were registered for Obama when he was a presidential candidate.[16] In terms of identifying specific Internet users or actors, formal names, while less specific, can also be used in conjunction with other IDs to track the contributions of specific users or Web 2.0 platform accounts, though an important caveat applies again because multiple techniques of identifying actors are often required when searching for networked campaigns and content across Web 2.0 platforms.

The last set of traffic tags we will discuss speaks more to the qualification and characterization of digital objects, a means by which posts, images, and videos are "tagged" using keywords, hashtags, and other content-related indices. Such user-generated forms of classification serve a central role in various projects that monitor trends on social media platforms such as Twitter or in the blogosphere, for instance, as aggregated by the platform-specific search engine technorati. Such tags particularly serve those working in the fields of information science, information retrieval, and library science to complicate objective means of classifying, controlling, and circulating documents and media objects. The emergence of a folksonomy epistemology, conversely, can also be overly celebrated as the ultimate freeing of information, wherein citizens not only produce and circulate their own political campaign objects, but also play a pivotal role in classifying their contributions to the networked political landscape.

Conclusions

While we recognize that this research has only begun to enumerate a new 2.0-inspired approach to the study of online networks and political networking, there are clear examples in the political sphere that suggest we are on the right path. Journalists now routinely track the origins of digital objects that seek to anonymously attack or parody public figures and politicians.[17] Such forms of political research are also practiced by political party staff. One of Canada's most social media–savvy reporters, for example, recently noted that opposition party staff in the Canadian capital matched the exact software code for a shade of blue used by the governing party on its party website (Pantone #333399) to a government website in an effort to argue that the current administration was politicizing—through similar branding—various governmental departments and programs.[18]

In this chapter, we have provided examples of code that can be analyzed to track political campaigns and communication across Web 2.0 platforms. However, much remains to be done. First, a road map of sorts is required to understand how—and under what conditions—a political party, blogger, or other user can best take advantage of the routes in, through, and across social media sites. Certain opportunities to network content between platforms are routinely prohibited. YouTube videos can be embedded on blogs, but until recently, not on Facebook or Twitter. Meanwhile, blog posts can be linked to Facebook friends' feeds but not to YouTube user accounts. Such distinctions are important to recognize when studying the effectiveness of online political campaigns, yet the speed at which such networked platforms emerge and then later change their back-end code and APIs makes such network mapping always and already out of date.

Studying political networking across Web 2.0 thus requires a commitment to experimenting with numerous traffic tags in order to track the uploading, spread, reuse, or remixing of various digital objects. Some sites provide for easy data collection with RSS feeds (such as blogs or information aggregators such as Google), while others such as Facebook and YouTube require an engagement with their API to collect large data sets. Again, just when one thinks that a sound method has been achieved to collect and track YouTube videos to blogs or Facebook, their API is changed, thus forcing researchers to readjust their methods again.

While the broader task of tracking relationships between platforms is fraught with pitfalls, the concept of traffic tags is fundamentally sound if one wants to understand the relationship between objects, users, and social media platforms. Shared 2.0 objects, such as Internet packets, need unique identifiers to distinguish themselves from each other. They provide the glue that binds together not only users (as "friends" on Facebook, for example) but also social media or 2.0 platforms. Without uploading, sharing, commenting, and remixing, there would be no networked media to map or exploit. Blog posts, comments, videos, and photos serve as molecular objects, always moving among the larger networked apparatus.

What is needed, then, is a road map that can point to how one can identify users and objects, and also how these can be tracked across social media platforms. This process requires constant updating to include new platforms, new functions, and new APIs. Such maps of traffic tags would consequently move research on political networks away from implied definitions of political connections or associations online, through an overreliance on hyperlink-mapping research, and on to a much richer understanding of what practices and sets of objects, actors, and 2.0 sites make for an effective (or botched!) networked campaign.

Notes

1. See, for example, the publication of in-camera party "talking points" on 2.0 sites such as Wikileaks. Available at http://michaelgeist.ca/content/view/3975/196/.
2. See http://www-958.ibm.com/software/data/cognos/manyeyes/.
3. For a list of representative software, see www.visualcomplexity.com.
4. For a representation of hyperlinks among political blogs in June 2008, see http://simoncollister.typepad.com/.shared/image.html?/photos/uncategorized/2008/06/26/polblogo.jpg.
5. Most notably, these objects include videos, digital images, blog posts, Twitter posts, and shared hyperlinks.
6. Cf. Hindman's *The Myth of Digital Democracy* (2009) for a good overview of http-based methods for network analysis.
7. The British Liberal Democrat Party (LDP) encodes a geo-tag in their http header that identifies their location as "Westminster, UK."
8 Available at http://www.conservative.ca, under the "viewpage source" option.
9. This tactic is often referred to as "meta-tag stuffing," and falls under the less subjective term "search-engine optimization." This topic has been vigorously debated by lawyers worldwide (McCarthy, 1999).
10. The most blatant example of so-called "meta-tag stuffing" refers to nefarious attempts to latch on to popular or trendy keywords used as search terms on Google to increase Internet traffic to websites—a form of traffic spam, if you will. The deregulated nature of meta-tag html-page encoding thus raises broad questions and concerns about the over-promotion of certain content (porn, dubious credit cards, etc.) and the burying of perhaps more socially relevant information. Ira S. Nathenson (1998) draws a clever yet frustrating analogy of a "spamdexed" network: "Imagine a never-ending traffic jam on a ten-lane highway. Road signs can't be trusted: the sign for Exit 7 leads to Exit 12, the sign for Cleveland leads to Erie. If you ask the guy at the Kwik-E-Mart how to get to I-79, he gives you directions to Route 30. To top it off, when you ask for a Coke, he gives you a Pepsi. Enough already. You stop at a pay phone to call directory assistance for the number to the local auto club, and instead get connected to 'Dial-a-porn'" (p. 45).
11. By using this biological term we mean to suggest that such http header tags and keywords

serve to implicate and reproduce both political languages and possible sites for articulating, networking, and organizing political agendas.

12. Since this search was conducted in July 2009, the results discussed here offer a significant "time-delayed" picture of the Conservative slogan; one that provides, perhaps, a more steeped view of the spread, adoption and reuse of the phrase.
13. See, for instance, the API test console on Facebook. Available at http://developers.facebook.com/tools.php.
14. The decision by the Obama administration to switch to an open-source Drupal website management suite was widely lauded by information activists. See http://drupal.org/node/375843.
15. Cf. http://www.mininova.org/user/aXXo for an online list of files uploaded by this "username." Wikipedia also provides an interesting overview of this "Internetalias"; see http://en.wikipedia.org/wiki/AXXo.
16. See http://inside.123-reg.co.uk/archives/domain-names-the-web-and-the-us-election.
17. Our own work with the Canadian Broadcasting Corporation (CBC) focused principally on determining the source of social media barbs and dirty tricks during the 2008 federal election in Canada. See http://www.cbc.ca/news/canadavotes/campaign2/ormiston/.
18. See http://blogs.canoe.ca/davidakin/main-page/the-colours-of-stephen-harpers-canada-that-would-be-pantone-333399/.

CHAPTER 7

Permanent Campaigning: A Mediatized Political Time and Space

This book has analyzed political events as a new mediated form of political communication, one defined increasingly through the technological plane as much as, if not more than, the rhetorical one. We have sought to clarify the concept of the permanent campaign beyond structuralist paradigms, whereby the terrain of politics is predominantly managed and controlled by a small group of professional politicians and party staff. From this older definition of permanent campaigning in the context of "new" media (see reference to Ornstein & Mann, 2000, in Chapter One), we have maintained the key notion of management of the political terrain as the effort to control not only the mediation of a political agenda, but more importantly, the mediatization of politics writ large (see Chapter Five). As such, this book has largely been concerned with how social media, shared digital objects, and new political actors define the very terrain upon which political communication is enabled. In sum, our book has been guided by this overarching question: what set of factors are forcing politics to adapt itself to the constraints, opportunities, possibilities, and limitations of existing media systems?

The Web 2.0 environment, characterized by the predominance of a handful of social media platforms, presents us with a new context of *mediatization*, one marked by the challenge of managing an ever-expanding field of communication in order to ensure the coherence, cohesiveness, and in everyday terms, *visibility* of political campaigns. In this regard, we have conceptualized and interrogated political campaigns as political moments which are themselves comprised of a set of issues, messages, and actions (*objects*), motivated by a set of *actors* that seek to manage possibilities of modes of communicative expression and actions (made manifest by social media *platforms*). Thus, to expand on our concern with the production of a mediatized political landscape, our book has also sought to understand how political campaigns in a social media environment maintained, or failed to maintain, their cohesiveness: how campaigns were, or were not, distorted or co-opted; how they were adapted to different types of reactions from political enemies, supporters, and publics. Hence, the networked concept of the permanent campaign developed in this book: the examination of the connections and disconnections between objects, platforms, and actors that enable or disable visibility, durability, coherence, and cohesiveness. The concept of networked permanent campaigning has thus allowed us to investigate the processes by which the coherence, durability, and visibility of a political campaign is realized in social media spheres, the mainstream media, and beyond.

The term *permanent campaigning* also helps us to avoid falling into the simplistic division of, on the one hand, a world of politics "as usual"—partisan bickering, agenda-setting, and media manipulations—and on the other hand, a world of new democratic politics, of previously disenfranchised populations armed with participatory technologies and in search of social and political change. Rather, if we are to understand the potential and limits of social media, then we have to pay more precise attention to how the social media context articulates the relationships between the different components of online permanent campaigns: the actors, platforms, and objects of the campaign.

Some might argue that our case studies were largely driven by electoral campaigns, which are quite distinct political moments characterized by intense and concerted efforts on the part of parties to manage political communication. However, our discussion of the Obama fund-raising campaign throughout the book's chapters, Chapter Two's discussion of the political activity of bloggers during political crises or outside of legislative frameworks, Chapter Three's investigation of political events that were born and articulated on Facebook, and even the comparative analysis of the weeks leading up the Australian federal election of 2007 sought to broaden the analysis of politics outside of the hyper-managed and increasingly predictable online election campaigns. Our goal has been to investigate changing practices of political actors, the quantity, quality, and context of objects (or issue-objects), and the effects of the predominance of particular social media platforms. In short, this book has sought to lay the groundwork for a broader understanding of a mediatized political culture: not just how Web-based campaigns are constructed to "win" elections, but how these campaigns are also designed to intervene on and across social media platforms to consistently raise questions about the credibility of particular politicians and leaders, opposition parties, alternative policies, and political solutions and agendas.

To speak of a permanent campaign is thus to raise questions about shifts and modulations in political time—the tempo of partisan life. Moreover, through the notion of a networked permanent campaign and a series of empirical case studies, we have questioned the transition from one political information cycle (the programmable 24-hour news cycle) to another that conforms more to a responsive political process—an immanent space of reaction to political events (see, in particular, our discussion of Twitter in Chapter Five). Political campaigns of any variety, electoral or otherwise, thus require an understanding of the fleeting nature of online communication. In Chapter Five we introduced the concept of compressed interface time as the software-mediated aspect of this fleeting form of communication. In so doing, we examined not just the limits of visibility on a screen (that has become increasingly smaller as tablets and handheld devices grow in popularity worldwide), but also the limited time before social media contributions such as comments on news websites and posts on Facebook and Twitter are buried in the unseen and sometimes unretrievable depths of the predominant information flow interface on many social media platforms—the vertical ticker.

Our notion of compressed interface-time in and on social media thus raises important questions, as noted above, about the management of the permanent campaign; that is, the efforts to cultivate any visible, sustained, and cohesive socially mediated—and mediatized—political campaign. Social media platforms, the unreliability—at times—of mobile partisans, and the visibility of socially media objects (videos, blog posts, comments, Twitter posts, etc.) likewise constantly shift their rules of engagement, their user conventions, and their governing code. Given these constantly shifting conventions, spaces, and subjects of social media, our book has often spoken with a 'methodological voice' in an effort to conceptualize a sustained study of a generalized set of logics, conventions, and technologies and their interactions with specific political moments. At the center of this approach is our triangular discussion of actors, objects, and platforms. Such an approach introduces important human variations and variables into an otherwise dauntingly technological landscape. Indeed, politics is still very much vetted and doled out by individuals and groups with goals and sometimes-coherent principles and agendas. In such a state of affairs, we have suggested that research begins by focusing on the articulation and the co-shaping of objects, platforms, and actors. None of these three poles—objects, platforms, and actors—exists independently from the others: actors are characterized by their engagement with platforms, objects are designed by actors and have to obey software protocols, social media platforms constantly work at increasing and maximizing their private logics of communication, sociability, social connection, and personalization.

The term *actors* is used in a purposely broad sense in order to avoid collapsing individuals into predefined political roles or ones that have been all too easily hybridized (citizen-journalist, media-blogger, etc.) and thus disconnected from their varying customs, conventions, temporalities, and spaces of communication and networking. Similarly, we found Schmidt's concept of the motorized or mobile partisan a helpful inspiration to understand partisan loyalty, proximity, and engagement in more centrally managed party campaigns, as explained in Chapters One and Two. The concept of a motorized partisan served as a relational one—again, not one that sought to define the blogger as, for example, a radically new political actor, or, conversely, a party mouthpiece. Rather, throughout the book we sought to integrate and define the "actor" as a contingent human component in a networked politics, one that develops new skills and literacies (learning the rules of the platforms) in an effort to highlight their issues-objects. We thus focused on actors as defined by their visible and traceable engagement in shaping, sharing, and sometimes distorting objects. Our intention was not to try to analyze what actors thought they were doing, or how they saw their roles in different political moments, but to focus on their performativity. We did this by studying not only the issues they engaged with but also the platforms they connected with and contributed to.

The mediated object (image, blog post, video, comment, etc.) is the political issue-object that is uploaded, shared, remixed, "liked," commented upon, downloaded, and so forth. Objects often unite different types of actors—they come to provoke debate, to comment, or, more proactively, to serve as a call to arms for other forms of political action and campaigning, from fund-raising to participating in rallies and protests, as we saw in Chapter Three. As such, objects cannot be understood in isolation; they are the handiwork of online actors who *perform* political communication, who are actively seeking not only to communicate but also to create an event around their objects, to sustain an object's visibility through social media conventions. The objects of the permanent campaign are mediatized: they have to be reshaped according to the conventions of the platforms, both at the interface level and at the code level. Some objects therefore become specific to platforms (i.e., a Facebook application), while others are shared across platforms, as in the case of embedded videos. In that sense, we also offered a conception of objects that moved away from their status as textual units and focused more on their networked characteristics—that is, on their trajectories on and across social media platforms. In tracking the paths of objects, we are able to get a clearer understanding of the reach of a political event online through time and space, and across media spheres. Understanding how different media spheres arc connected through shared objects, or else disconnected from one another at a particular moment because there have no object in common, helps us understand the relationships between actors and how specific groups avoid each other (see Chapter Three on Facebook publics) or, on the contrary, build temporary or more permanent alliances (Chapter Two). The tracing of the distribution of objects, as noted in Chapter Six in particular, is increasingly more challenging to conduct as we are faced with proprietary types of platforms and the multiplication of codes, languages, and protocols.

Our decision to analyze objects as mediatized and networked sought to introduce a central mediator of both politics (the content of debate and discourse that often embodies or signifies particular issues) and technology (the discrete, *digital* entity): the platform. Objects, of course, are the lifeblood of social media platforms, from the most aggressively proprietorial to the intensely anarchic and chaotic. While increasing user accounts is a central measure of success for social media platforms, their modus operandi continues to center on the active uploading, sharing, and circulation of digitized objects—be it audio, text, still image, or video. Since social media platforms have increasingly sought to develop platform-specific rules and algorithms to manage the visibility of objects, whether in a more public sense, as in the case of Google's Australia Votes portal, or in a more personalized sense, as was the case with Facebook, we argued in Chapter Six for an understanding of how actors motivate and interact with issue-objects, and how such objects must conform to the rules of specific social media platforms. As a consequence, social media cannot be understood as a common set of opportunities and functions; rather, through a cross-platform lens, we have argued that social media platforms serve to strategically connect—and disconnect—possible forms of visibility of issue-objects, and conventions for partisan actors to interact and engage with said issue-objects.

Moving forward, we suggest that there needs to be additional work on how less institutionalized, and also more international, political campaigns, protests, occupations and the like, can seek to mobilize their own set of actors, objects, and platforms to counter and compete with governments, political parties, and corporate actors who are typically in a better resourced position to deploy permanent campaign-like tactics. NGOs, community groups, activists, and others often develop and deploy their own distinct political campaigns, in the streets and online—campaigns that broadly speak to the ideologies, goals, and policies of each respective group. From time to time, though, groups come together to join forces and conduct joint campaigns—often in an effort to change government policies, and of course, in advance of and during election campaigns. It is here that the permanent campaign might offer new opportunities for civil society.

In a political culture dominated by a permanent campaign, we argue that such groups—working in much more subtle, distributed, and yes, "permanent" coalitions—might develop intermittent and cumulative forms of contemporary political communications and campaigning; campaigns that are made common by—and through—the shared and/or similarly designed (aesthetically, digitally, or otherwise) and intermittently circulated issue-objects. Such a common, intermittent, object-centered and networked "campaign," drawing upon a persistent mimetic-like strategy, would serve as a constant and immanent reminder of contested political issues, processes, and cultures—reminders that could be made yet more powerful through attention and amplification by established political actors, and the mainstream media—during heightened periods of political activity.

The permanent campaign thus should not be defined or dismissed as a political control technology. Rather, the permanent campaign is a contested plane, a set of conventions, possibilities, and limitations—not one where established forms of political power (politicians, celebrities, etc.) are simply exposed or at times embarrassed, but where they are consistently called into question, contradicted, and potentially destabilized, as we have seen in some jurisdictions around the world, through the uploading, downloading, sharing, remixing, and embedding of issue-objects.

Index

General Editor: Steve Jones

Digital Formations is the best source for critical, well-written books about digital technologies and modern life. Books in the series break new ground by emphasizing multiple methodological and theoretical approaches to deeply probe the formation and reformation of lived experience as it is refracted through digital interaction. Each volume in **Digital Formations** pushes forward our understanding of the intersections, and corresponding implications, between digital technologies and everyday life. The series examines broad issues in realms such as digital culture, electronic commerce, law, politics and governance, gender, the Internet, race, art, health and medicine, and education. The series emphasizes critical studies in the context of emergent and existing digital technologies.

Other recent titles include:

Felicia Wu Song
Virtual Communities: Bowling Alone, Online Together

Edited by Sharon Kleinman
The Culture of Efficiency: Technology in Everyday Life

Edward Lee Lamoureux, Steven L. Baron, & Claire Stewart
Intellectual Property Law and Interactive Media: Free for a Fee

Edited by Adrienne Russell & Nabil Echchaibi
International Blogging: Identity, Politics and Networked Publics

Edited by Don Heider
Living Virtually: Researching New Worlds

Edited by Judith Burnett, Peter Senker & Kathy Walker
The Myths of Technology: Innovation and Inequality

Edited by Knut Lundby
Digital Storytelling, Mediatized Stories: Self-representations in New Media

Theresa M. Senft
Camgirls: Celebrity and Community in the Age of Social Networks

Edited by Chris Paterson & David Domingo
Making Online News: The Ethnography of New Media Production

To order other books in this series please contact our Customer Service Department:

(800) 770-LANG (within the US)
(212) 647-7706 (outside the US)
(212) 647-7707 FAX

To find out more about the series or browse a full list of titles, please visit our website:

WWW.PETERLANG.COM